Remote Sensing of Hydro-Meteorology

Remote Sensing of Hydro-Meteorology

Editors

Joo-Heon Lee
Jong-Suk Kim
Young Hun Jung
Chanyang Sur

MDPI • Basel • Beijing • Wuhan • Barcelona • Belgrade • Manchester • Tokyo • Cluj • Tianjin

Editors

Joo-Heon Lee
Department of Civil Engineering
Joongbu University
Goyang
Korea

Jong-Suk Kim
Department of Hydrology and
Water Resources
Wuhan University
Wuhan
China

Young Hun Jung
Construction and Disaster
Prevention Engineering
Kyungpook National University
Sangju
Korea

Chanyang Sur
Department of Civil and
Environmental Engineering
Sejong University
Seoul
Korea

Editorial Office
MDPI
St. Alban-Anlage 66
4052 Basel, Switzerland

This is a reprint of articles from the Special Issue published online in the open access journal *Remote Sensing* (ISSN 2072-4292) (available at: www.mdpi.com/journal/remotesensing/special_issues/hydro_meteorology).

For citation purposes, cite each article independently as indicated on the article page online and as indicated below:

LastName, A.A.; LastName, B.B.; LastName, C.C. Article Title. *Journal Name* **Year**, *Volume Number*, Page Range.

ISBN 978-3-0365-1630-1 (Hbk)
ISBN 978-3-0365-1629-5 (PDF)

© 2022 by the authors. Articles in this book are Open Access and distributed under the Creative Commons Attribution (CC BY) license, which allows users to download, copy and build upon published articles, as long as the author and publisher are properly credited, which ensures maximum dissemination and a wider impact of our publications.

The book as a whole is distributed by MDPI under the terms and conditions of the Creative Commons license CC BY-NC-ND.

Contents

About the Editors vii

Preface to "Remote Sensing of Hydro-Meteorology" ix

Jong-Suk Kim, Seo-Yeon Park, Joo-Heon Lee, Jie Chen, Si Chen and Tae-Woong Kim
Integrated Drought Monitoring and Evaluation through Multi-Sensor Satellite-Based Statistical Simulation
Reprinted from: *Remote Sens.* **2021**, *13*, 272, doi:10.3390/rs13020272 1

Jong-Suk Kim, Anxiang Chen, Junghwan Lee, Il-Ju Moon and Young-Il Moon
Statistical Prediction of Typhoon-Induced Rainfall over China Using Historical Rainfall, Tracks, and Intensity of Typhoon in the Western North Pacific
Reprinted from: *Remote Sens.* **2020**, *12*, 4133, doi:10.3390/rs12244133 19

Jong-Suk Kim, Sun-Kwon Yoon, Sang-Myeong Oh and Hua Chen
Seasonal Precipitation Variability and Non-Stationarity Based on the Evolution Pattern of the Indian Ocean Dipole over the East Asia Region
Reprinted from: *Remote Sens.* **2021**, *13*, 1806, doi:10.3390/rs13091806 33

Jong-Suk Kim, Phetlamphanh Xaiyaseng, Lihua Xiong, Sun-Kwon Yoon and Taesam Lee
Remote Sensing-Based Rainfall Variability for Warming and Cooling in Indo-Pacific Ocean with Intentional Statistical Simulations
Reprinted from: *Remote Sens.* **2020**, *12*, 1458, doi:10.3390/rs12091458 51

Jie Chen, Ziyi Li, Lu Li, Jialing Wang, Wenyan Qi, Chong-Yu Xu and Jong-Suk Kim
Evaluation of Multi-Satellite Precipitation Datasets and Their Error Propagation in Hydrological Modeling in a Monsoon-Prone Region
Reprinted from: *Remote Sens.* **2020**, *12*, 3550, doi:10.3390/rs12213550 69

Hamid Mohebzadeh, Junho Yeom and Taesam Lee
Spatial Downscaling of MODIS Chlorophyll-a with Genetic Programming in South Korea
Reprinted from: *Remote Sens.* **2020**, *12*, 1412, doi:10.3390/rs12091412 103

Shanlei Sun, Jiazhi Wang, Wanrong Shi, Rongfan Chai and Guojie Wang
Capacity of the PERSIANN-CDR Product in Detecting Extreme Precipitation over Huai River Basin, China
Reprinted from: *Remote Sens.* **2021**, *13*, 1747, doi:10.3390/rs13091747 123

About the Editors

Joo-Heon Lee

Prof. Joo-Heon Lee currently works at the Civil Engineering Department, Joongbu University. Joo-Heon Lee does research in Drought Monitoring & Forecasting, Remote Sensing, Climatology, and Eco Hydrology. Their current projects are Change Adaptation in Water Resources in Koreaand Drought Impact Assessment

Jong-Suk Kim

Prof. Jong-Suk Kim works at the Department of Hydrology and Water Resources, Wuhan University, China. His research work focuses on diagnostic analysis and assessment methodologies to understand hydroclimatic change, nonstationarity in hydroclimatic extremes, and an integrated view of climate and hydrologic variability and changes.

Young Hun Jung

Prof. Younghun Jung currently works at the Construction and Disaster Prevention Engineering, Kyungpook National University, Sangju. Younghun does research in Remote Sensing, Hydrology, and Environmental Engineering. Their most recent publication is -Based BFlow System for the Assessment of Streamflow Characteristics at National Level.

Chanyang Sur

Dr. Sur currently works at the Department of Civil and Environmental Engineering, Sejong University in Korea. His research focuses on land surface–atmosphere interactions, soil moisture analysis, remote sensing, and climate change.

Preface to "Remote Sensing of Hydro-Meteorology"

Extreme hydrometeorological events that occur naturally threaten and cause harm to lives and livelihoods and result in billions of dollars of damage worldwide every year. Their environmental impacts are equally catastrophic. Human activities may help prevent hydrometeorological hazards from turning into disasters, but in many situations, they may also exacerbate their impacts, e.g., through excessive development in coastal areas that increase risk exposure and community vulnerability. Moreover, climate change may be responsible for the increasing frequency and magnitude of atmospheric patterns that lead to more frequent and intense hydrometeorological disasters (e.g., severe storms, floods, and droughts).

This book focused on remote sensing of hydrologic extremes, including flood/drought monitoring, risk management and policy, hydrometeorological extremes and their impact on human-environment systems, frequency analysis, and vulnerability assessment for adaptation to climate change.

Joo-Heon Lee, Jong-Suk Kim, Young Hun Jung, Chanyang Sur
Editors

Article

Integrated Drought Monitoring and Evaluation through Multi-Sensor Satellite-Based Statistical Simulation

Jong-Suk Kim [1], Seo-Yeon Park [2,*], Joo-Heon Lee [2], Jie Chen [1], Si Chen [3] and Tae-Woong Kim [4]

1. State Key Laboratory of Water Resources and Hydropower Engineering Science, Wuhan University, Wuhan 430072, China; jongsuk@whu.edu.cn (J.-S.K.); jiechen@whu.edu.cn (J.C.)
2. Department of Civil Engineering, Joongbu University, Gyeonggi-do 10279, Korea; leejh@joongbu.ac.kr
3. School of Resources and Environment, Hubei University, Wuhan 430062, China; kathryncs123@hotmail.com
4. Department of Civil and Environmental Engineering, Hanyang University (ERICA), Gyeonggi-do 15588, Korea; twkim72@hanyang.ac.kr
* Correspondence: sypark276@gmail.com

Abstract: To proactively respond to changes in droughts, technologies are needed to properly diagnose and predict the magnitude of droughts. Drought monitoring using satellite data is essential when local hydrogeological information is not available. The characteristics of meteorological, agricultural, and hydrological droughts can be monitored with an accurate spatial resolution. In this study, a remote sensing-based integrated drought index was extracted from 849 sub-basins in Korea's five major river basins using multi-sensor collaborative approaches and multivariate dimensional reduction models that were calculated using monthly satellite data from 2001 to 2019. Droughts that occurred in 2001 and 2014, which are representative years of severe drought since the 2000s, were evaluated using the integrated drought index. The Bayesian principal component analysis (BPCA)-based integrated drought index proposed in this study was analyzed to reflect the timing, severity, and evolutionary pattern of meteorological, agricultural, and hydrological droughts, thereby enabling a comprehensive delivery of drought information.

Keywords: remote sensing; integrated drought monitoring; meteorological drought; hydrological drought; agricultural drought; Bayesian principal component analysis (BPCA); statistical simulation

1. Introduction

Droughts, along with floods, are some of the most common and inevitable natural disasters faced by human beings [1–4]. Therefore, many researchers have been trying to monitor and predict droughts accurately, and the development of drought monitoring techniques based on satellite remote sensing (RS) data (as a representative method) has garnered special interest in recent years [4–9]. The onset and magnitude of drought in the region is still a challenge for researchers because of a lack of ground meteorological observatories [4]. However, satellite-based RS data partially solve the problem by providing information in a fast and cost-effective way. The advantage of RS-based monitoring using satellite data is that it is possible to monitor droughts in large areas and ungauged basins, and we can utilize multiple satellite imagery data to have accurate results; therefore, monitoring drought by using satellites has proven to be an efficient and reliable tool [6,10–13].

There are four kinds of droughts in the academic sense: meteorological, agricultural, hydrological droughts, and their socioeconomic impacts [2,14–16]. A meteorological drought is caused by a deficit through the shortage of rainfall and is mainly a short-term drought event [6]. An agricultural drought is determined based on the vitality of vegetation and the pattern of quantitative changes in soil moisture; it indicates short or medium-term drought situations [6,13]. A hydrological drought is commonly a mid or long-term drought condition; this drought identification is made based on a shortage of water resources required by human-environmental systems, such as river discharge, efficient water levels of dams, and reservoir storage [17]. A socioeconomic drought consists of a wide range that takes meteorological, agricultural, and hydrological droughts into account and is characterized by the temporal and spatial processes of water demand and supply [18].

A variety of drought indicators that help prevent disasters and reduce and allocate water resources have been developed to quantify different drought conditions, such as severity, duration, and frequency [3,5,17,19–25]. The standardized precipitation index (SPI; [19]) and the Palmer drought severity index (PDSI; [26]) are the most commonly used meteorological drought indices. The SPI standardization concept was also applied to other drought indices, such as the standardized runoff index (SRI; [20]) and standardized soil moisture index (SSI [22]).

Because RS technology provides an alternative approach for analyzing drought events across a wide range of regions, many studies have introduced RS-based drought indices [5,7,23,24,27]. Zhang and Jia [27] proposed the microwave integrated drought index (MIDI) to monitor meteorological drought over semi-arid regions and the continental United States of America. Cunha et al. [23] calculated the normalized differences vegetation index (NDVI) and land surface temperature (LST) data to monitor the effects of drought on vegetation in real-time. Sur et al. [24] analyzed Korea's drought conditions through a comparative analysis of the existing drought indices (SPI and PDSI) based on a satellite image-based drought index from 2004 to 2013. It was confirmed that the results of the evaporative stress index (ESI), and the energy-based water definition index (EWDI) showed high applicability for severe drought situations since 2010. Cong et al. [5] selected three widely used satellite drought indices as indicators suitable for drought monitoring in northeastern China and investigated the spatiotemporal patterns and trends of rainfall and drought; the indices were normalized monthly precipitation anomaly percentage (NPA), vegetation health index (VHI), and normalized vegetation supply water index (NVSWI). Zhang et al. [28] combined the global land data assimilation system version 2 (GLDAS-2) soil moisture data and NDVI with crop phenology data and assessed drought evolution and crop growth. Sur et al. [7] developed a new agricultural drought index called the agricultural dry condition index (ADCI) by combining various hydrometeorological variables and verified the applicability of the ADCI on the yield of paddy and arable crops in Korea.

Through the review of previous studies, we can assume that information can be integrated from multi-sensor satellite data and multivariate analyses to effectively achieve comprehensive drought assessment goals. In addition to providing information based on different drought conditions (meteorological, agricultural, and hydrological), it is necessary to develop and apply an integrated drought index that considers complex factors that can provide comprehensive information about droughts and the required proactive response to drought situations. Inspired by this idea, our study seeks to diagnose complex droughts by using multi-sensor collaborative approaches and multivariate dimensional reduction models. In this study, we proposed an integrated drought assessment method to comprehensively convey drought information to the public and conducted statistical simulations to determine spatial sensitivity to various types of droughts to provide tailored information on local drought responses in a changing climate.

2. Materials and Methods

2.1. Multi-Sensor Drought Indices

2.1.1. Standardized Precipitation Index (SPI)

The SPI is a drought index developed with the idea that it is initiated by a decrease in precipitation, thereby causing water shortage (compared to the relative water demand). In other words, it was developed from the above assumption that decreased precipitation has different effects on groundwater, reservoir storage, soil moisture, and river runoff. The SPI is an efficient way to calculate the impact of individual water sources on droughts by setting time units accumulated over a given period of time (over 1, 3, 6, and 12 months), and calculating the drought index by using the amount of precipitation on a time basis [19,28].

The SPI is also recommended by the World Meteorological Organization (WMO) for tracking meteorological droughts [21,25].

2.1.2. Agricultural Dry Condition Index (ADCI)

The ADCI is an agricultural drought index that takes into account the vegetation conditions, soil moisture, and LST of the affected region. First, the vegetation condition index (VCI) is applied for vegetation analysis. Sur et al. [7] proposed the ADCI as a new agricultural drought index, which is a combination of the three indices mentioned above (SMSI, VCI, and TCI). The cause of the agricultural drought was developed based on the concept of reducing the vitality of vegetation due to the lack of soil moisture and overheating of the surface temperature caused by high temperatures, developing into agricultural drought as this phenomenon continues. The ADCI can be calculated by using the Equation (1) given below:

$$\text{ADCI} = 0.6 * SMSI + 0.2 * VCI + 0.2 * TCI. \tag{1}$$

The VCI is a suitable index for agricultural drought monitoring, such as temporal and spatial vegetation changes and the onset and intensity of drought [29,30]. Kogan [29] developed the VCI, which was standardized using the maximum and minimum values of the NDVI developed based on the notion that droughts do not provide normal water supply to plants (Equation (2)).

$$\text{VCI} = \frac{NDVI - NDVI_{min}}{NDVI_{max} - NDVI_{min}}, \tag{2}$$

where $NDVI_{min}$ and $NDVI_{max}$ represent the minimum and maximum values of NDVI for the entire period of the pixel. The following is the LST-related index, called the temperature condition index (TCI), which is an index developed by Kogan [29] based on the fact that LST affects the stress of vegetation and is one of the drought factors that affect soil moisture. The TCI is a standard LST that uses the maximum and minimum LST values and as shown in the following equation (Equation (3)).

$$\text{TCI} = \frac{LST_{max} - LST}{LST_{max} - LST_{min}} \tag{3}$$

Finally, soil moisture needs to be considered for determining the ADCI. The soil moisture saturation index (SMSI) assumes that soil moisture is directly proportional to thermal inertia (TI). One of TI's simple approximations is the apparent thermal inertia (ATI), which can be derived in Equation (4); note that we assume that the solar energy is uniform.

$$\text{ATI} = \frac{(1 - \alpha)}{LST_{day} - LST_{night}}, \tag{4}$$

where α is the land surface albedo and LST_{day} and LST_{night} are the surface temperatures during day and night, respectively. The SMSI can be calculated using the ATI, as shown in the following equation (Equation (5)).

$$\text{SMSI} = \frac{ATI - ATI_{min}}{ATI_{max} - ATI_{min}} \tag{5}$$

2.1.3. Water Budget-Based Drought Index (WBDI)

The water budget-based drought index (WBDI), which was proposed by Sur et al. [18], was developed by adopting the water balance perspective and by using precipitation and evaporation as input variables. The evaporation of water balance is caused primarily by changing the state of water, which is achieved by changing the temperature [31]. The WBDI is defined as the difference between precipitation and evaporation as surface runoff and sub-surface runoff in the water budget equation, as given below (Equation (6)):

$$P - E = dS + R, \qquad (6)$$

where P is the precipitation (mm), E is actual evaporation (mm), dS is soil moisture change (mm), and R is the potential runoff (mm). The above results are treated as possible runoff in the basin and expressed in an index, as given below (Equation (6)):

$$WBDI = z(P - E), \qquad (7)$$

where z denotes the standardization. Instead of monitoring the current precipitation and drought conditions through evaporation, the WBDI, estimated by using the water balance formula, defines a hydrological drought through potential (near future) runoff, thereby adopting a short-term prognosis approach.

2.2. Study Area and Remote Sensing Data

In this study, we used the moderate resolution imaging spectroradiometer (MODIS), precipitation estimation from remotely sensed information using an artificial neural network climate data record (PERSIANN-CDR), and global precipitation measurement (GPM) integrated multi-satellite retrievals for GPM (GPM IMERG) to calculate various drought indices. Through the MODIS satellite, the LST (MOD11A1), NDVI (MOD13A3), actual evapotranspiration (AET; MOD16A2), and albedo (MCD43B2) data from 2001 to 2019 were collected (Table 1). To obtain the precipitation data, we used the PERSIANN-CDR data from 1983 to 1997 that was generated by the center for hydrometeorology and remote sensing (CHRS) at the University of California in Irvine; the data were obtained before the tropical rainfall measuring mission (TRMM). The TRMM data from 1998 were utilized, and among many data, the gridded TRMM3B42 data were collected until 2014 (at the end of TRMM's life), which was provided by the National Aeronautics and Space Administration (NASA) [32]. Following 2014, we used data from GPM IMERG that obtained data until 2019 to calculate the meteorological drought index [33]. Among the GPM IMERG data, the data after the last four months of the calibration were used to enhance the reliability of the precipitation data. Due to the different spatial and time resolutions of the collected data, the spatial resolution was set at 1×1 km and the time resolution was considered to be monthly, which is consistently reprocessed. The main areas of this study were the five major rivers of the Korean Peninsula, and we analyzed 849 sub-basins (Figure 1).

Table 1. Remote sensing (RS) data used in this study.

	Product		Resolution	Data Period
MODIS	MOD11A1	Land Surface Temperature	1 km, daily	2001–2019
	MOD13A3	Vegetation Indices	1 km, monthly	
	MOD16A2	Evapotranspiration	0.5 km, 8 days	
	MCD43B2	Albedo	1 km, 8 days	
PERSIANN-CDR	PERSIANN-CDR	Precipitation	25°, daily	1983–1997
TRMM	TRMM3B42	Precipitation	25°, 3 h	1998–2014
GPM	GPM IMERG	Precipitation	10°, 30 min	2015–2019

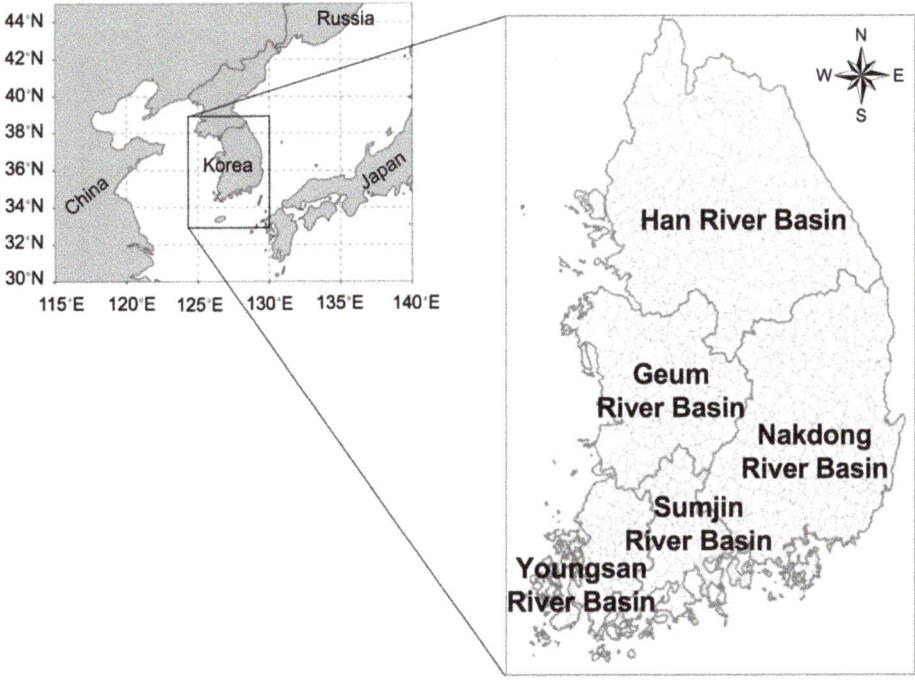

Figure 1. Geographical location of the five major river basins and the 849 sub-basins in Korea.

2.3. Integrated Drought Monitoring with Multi-Sensor Based Statistical Simulations

The types of RS data, mainly used for drought monitoring, depend on the type of satellite used; however, data such as precipitation, vegetation, surface temperature, soil moisture, and evaporation are mostly used. The data can be used individually. However, drought phenomena may not be sufficient for drought analysis based on a single indicator because it is related to a number of variables [34]. However, it may be more useful to combine information in the form of an appropriate drought index for more accurate monitoring of complex drought phenomena.

Hao and AghKouchak [23,34] proposed the multivariate standardized drought index (MSDI) based on a copula distribution or nonparametric joint distribution for a bivariate model of precipitation and soil moisture. However, with recent advances in technology, the size and complexity of data tend to increase. Such complexity makes it difficult to detect the dependence between the response variable and covariates because of the enormous number of available covariates [35]. To resolve these problems, an approach to reducing the number of covariates (through dimension reduction) is being used. Principal component analysis (PCA) is a tool that is commonly used for dimension reduction [36] and is a feature transformation method that directly transforms the variables (in dimension reduction methods) without losing much of the data's inherent attribution information. In this study, for the three different multi-variables acquired from the satellite data, an integrated drought index is calculated through the application of the Bayesian PCA (BPCA; [37,38]) and intentionally biased bootstrap (IBB; [39]) simulation for characterizing three aspects of meteorological (using SPI), agricultural (using ADCI), and hydrological (using WBDI) droughts. The BPCA approach can estimate the intrinsic dimensionality of the multi-dimensional dataset with missing data, which is suitable for application to satellite data, and has been evaluated as an accurate and robust model [37,38,40]. The BPCA analysis is performed by using the following three procedures: principal component (PC) regression, Bayesian estimation, and an expectation-maximization (EM)-like repetitive algorithm [38].

The IBB applied for statistical simulation of drought indices is a kind of weighted bootstrap that follows constraints that are designed to select resampling probabilities and conditionally applied to data; this helps to improve the statistical performance and minimizes the distance of weighted distributions [39,41]. This study employed the IBB to evaluate regional drought changes in meteorological (SPI), agricultural (ADCI), and hydrological (WBDI) droughts and analyzed the relative sensitivity of each drought index to the RS-based integrated drought index (RSIDI) calculated by using the BPCA (Figure 2). The IBB applied in this study is described as follows.

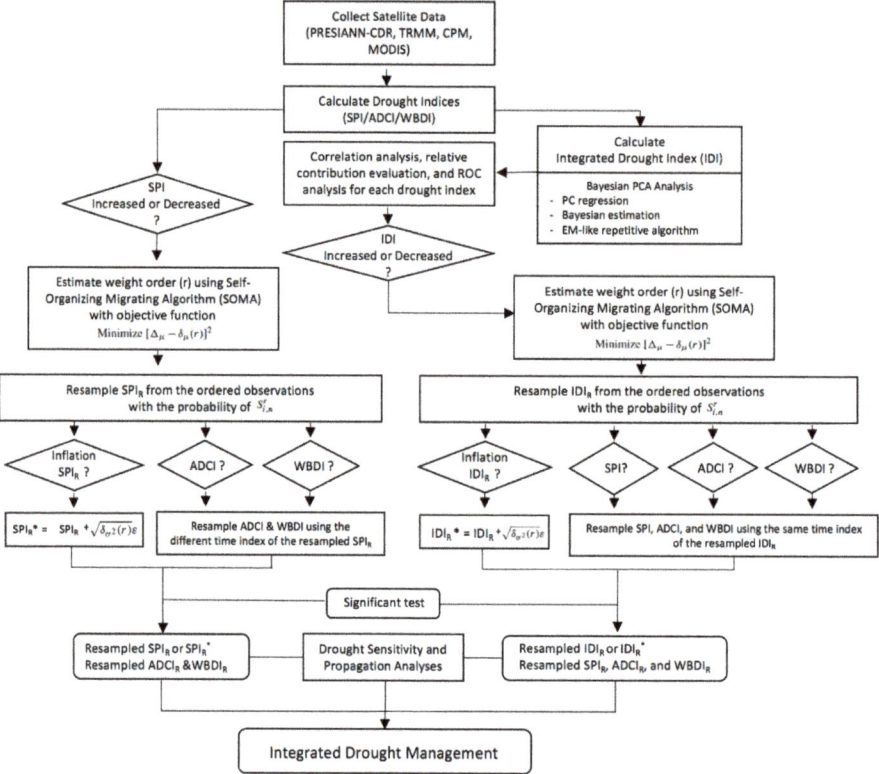

Figure 2. Procedure of intentionally biased bootstrap (IBB) analysis for integrated drought management.

The IBB simulation re-samples the observations X_i to n replacement (e.g., bootstrapping) by intentionally increasing or decreasing the data by as much as δ_μ. The data X_i are increasingly ordered by assigning different weights $W_{i,n}$ according to the magnitudes of the observations, as given below:

$$W_{i,n} = i/n, \qquad (8)$$

where $i = 1, 2, 3, \ldots, n$, and the data matrix is rearranged in the same order as the ordered X_i. The assigned weight $W_{i,n}$ represents the probability of selection for X_i data in the IBB simulation. The intentional change of increase or decrease (δ_μ) can be calculated as given in Equation (9).

$$\delta_\mu = \tilde{\mu} - \hat{\mu} = \frac{1}{\psi}\sum_{i=1}^{n} W_{i,n} X_i - \frac{1}{n}\sum_{i=1}^{n} X_i, \qquad (9)$$

$$\psi = \sum_{i=1}^{n} W_{i,n}. \qquad (10)$$

Equation (9) can be generalized with a weight order (r) as Equation (11).

$$\delta_\mu(r) = \widetilde{\mu}(r) - \hat{\mu} = \frac{1}{\psi_r} \sum_{i=1}^{n} W_{i,n}^r X_i - \frac{1}{n} \sum_{i=1}^{n} X_i \quad (11)$$

The selection of the weight order (r) can be performed by using the self-organizing migrating algorithm (SOMA; [42]) with the following objective function (Equation (12)):

$$minimize \ [\delta_\mu - \delta_\mu(r)]^2. \quad (12)$$

Note that if $r < 0$, then $\delta_\mu(r) < 0$, which implies a drier state, and if $r > 0$, then $\delta_\mu(r) > 0$, which implies a wetter state. When $r < 0$, lower values indicating the dry state are resampled more frequently than higher values indicating the humid state, causing $\delta_\mu(r)$ to decrease. In addition, to objectively determine the accuracy of drought monitoring using three satellite-based drought indices, this study conducted a receiver operation characteristics (ROC) analysis. The ROC analysis was performed to evaluate the validity of the RSIDI calculations using the three drought indices (SPI, ADCI, and WBDI). The range of drought indices used in this study is given in Table 2.

Table 2. Range of the drought indices used in this study.

Drought Condition	SPI	ADCI	WBDI	RSIDI
Normal	>0	>40	>0	>0
Attention	−1.0–0	30–40	0––0.5	−1.0–0
Caution	−1.0––1.5	20–30	−0.5––1.0	−1.0––1.5
Alert	−1.5––2.0	10–20	−1.0––1.5	−1.5––2.0
Serious	<−2.0	0~10	<−1.5	<−2.0

3. Results

In this study, the RSIDI was extracted for 849 sub-basins over the five major Korean river basins using a BPCA-based combination model for the three drought indices for 2001–2019. As a result of the evaluation of the proportion of variation (POV) of the three drought indices by region, the BPCA-based RSIDI explained the average POV (68.9%) of the 849 sub-basins (Han River basin: 68.1%, Nakdong River basin: 68.7%, Geum River basin: 71.3%, Youngsang River basin: 71.7%, and Sumjin River basin: 71.0%), showing a high POV, especially in the southern part of the country. In addition, the calculated RSIDI showed a relatively high correlation with SPI (median: 0.96) and WBDI (median: 0.96). In the case of the ADCI (maximum: 0.91, median: 0.53), the correlation with RSIDI was broad in the region and showed a relatively weak correlation in some areas of the Han and Nadkong River basins; however, on an average, the correlation was 0.53 (p-value < 0.001) in 849 sub-basins, indicating that the RSIDI can provide robust and comprehensive integrated drought information by maintaining the inherent characteristics of the three drought indices (Figure 3). By resolving the system equations for each drought index, it has been shown that the relative contribution of the RSIDI to each time-series can be assessed. Figure 4 illustrates the relative contribution of each drought index to the Gojicheon stream (#10011) of the Han River basin.

In the following section, the droughts in 2001 and 2014, which are the representative severe drought years of severe droughts that have occurred since the 2000s, were assessed using the RSIDI produced by multi-sensor satellite data and multivariate analysis. The application of integrated drought monitoring based on satellite data was evaluated through spatial-temporal variability analysis between the RSIDI and other drought indices using the ROC analysis to test the accuracy of the models. In addition, the onset, intensity, and evolution patterns of droughts were compared to each drought index, and the applicability of the RSIDI was evaluated through an IBB simulation.

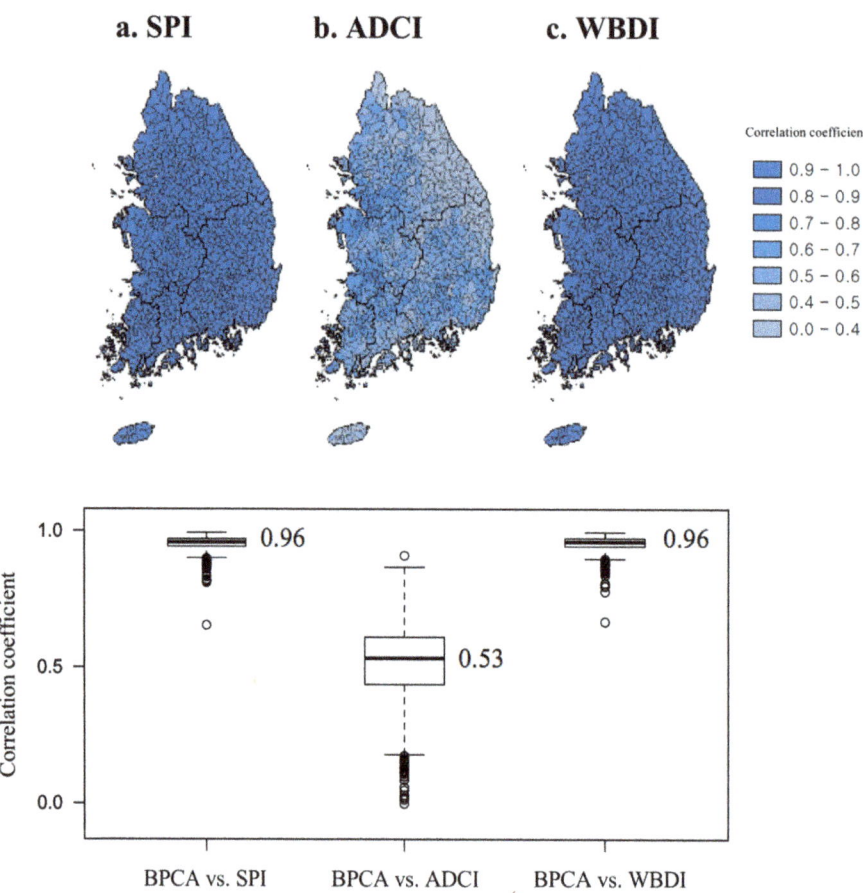

Figure 3. Results of correlation analysis with RS-based integrated drought index (RSIDI) through Bayesian PCA (BPCA); (**a**) SPI (standardized precipitation index), (**b**) ADCI (agricultural dry condition index), and (**c**) WBDI (water budget-based drought index). The lower panel in the figure results from the analysis of the correlation of 849 sub-basins summarized in the boxplot and is illustrated by each applied drought index namely, SPI, ADCI, and WBDI.

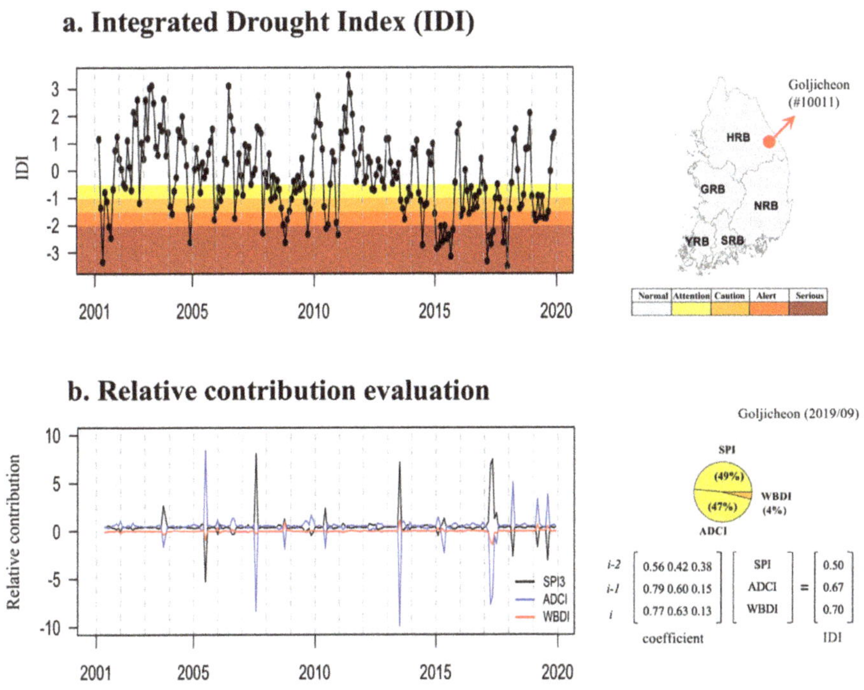

Figure 4. Time series of integrated drought index (IDI) for 2001–2019 in the Gojicheon stream in the Han River basin. (**a**) IDI index and (**b**) relative contribution evaluation.

3.1. Drought Impact Assessment and Drought Monitoring

The RSIDI was evaluated for the 2001 drought (Figures 5 and 6). The severe spring drought in 2001 began in the fall of 2000 and lasted until the spring of 2001. In spring, when the agricultural water demand was the highest due to the rice planting, the supply of agricultural water was insufficient, causing serious agricultural damages. In most parts of the Korean Peninsula, less than 50% of the average annual rainfall was recorded, and in some areas, only 10 to 30% of the average annual rainfall resulted in the most extensive agricultural drought damage in June [43]. The drought that occurred in 2001 was mostly resolved after more than 150 mm of rainfall since mid-June.

The SPI and WBDI illustrate the drought from April to May was a serious event, and the drought centered in the central region since September also appears to be approaching the serious stage (Figure 5). The ROC analysis between the RSIDI and three drought indices also showed that the WBDI had the highest with 0.90, followed by SPI at 0.78, and ADCI at 0.65. For ADCI, the observed indicators showed significant spatial variation compared to the other drought indices, and in the 2000 drought, the effects of changes in the SMSI resulted in a weaker or earlier drought peak than those observed in cases of other drought indices (Figure 6). These features appear to be more sensitive to short-term droughts as the ADCI applied in this study was used only for the surface soil moisture.

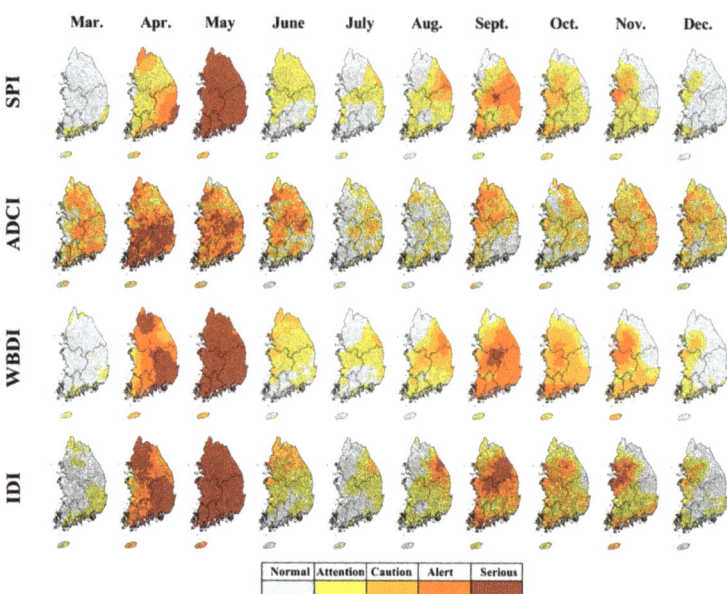

Figure 5. Spatial change of each drought index: SPI, ADCI, WBDI, and IDI (integrated drought index) for the 2001 drought.

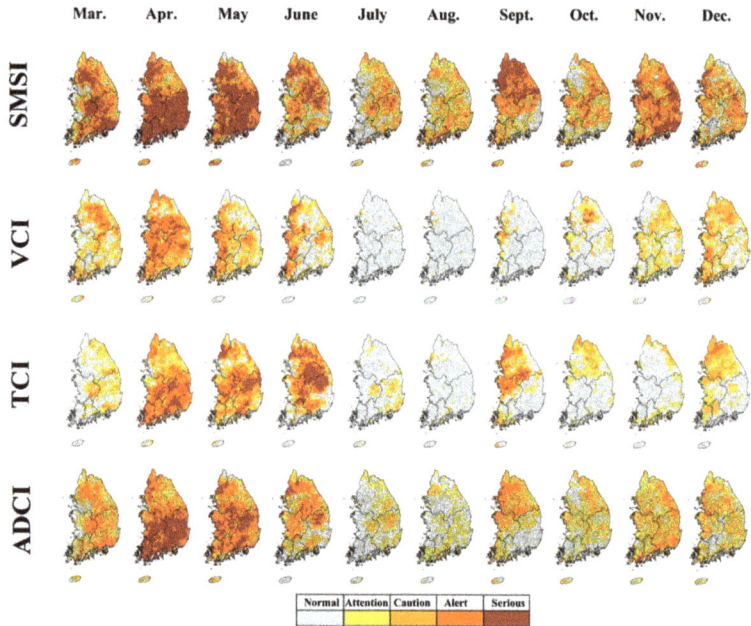

Figure 6. Spatial change of each drought index: SMSI (soil moisture saturation index), VCI (vegetation condition index), TCI (temperature condition index), and ADCI for the 2001 drought.

Figures 7 and 8 show the results of the evaluation of the 2014 drought. In 2014, a drought occurred around the Han River basin; in the same year, the Gangwon Province had 70% of the average annual rainfall and the Gyeonggi Province area around Seoul had 59%, which was less compared to the previous years' average annual rainfall. In particular, in 2014, a dry monsoon phenomenon occurred in which a drought that began in spring (due to the El Niño phenomenon) continued to cause no rainfall or a significantly lower amount of rainfall [44]. The rainfall between June 2014 and July 2014 was 48% (compared to the average in the past years), and the national water storage level also dropped significantly to 64% over the previous year. In August, the average water level of multi-purpose dams in Korea was only 36.1%. The droughts lasted until 2015, resulting in less than 70% of the average rainfall, and the hydrological droughts, with reservoirs in many multipurpose dams, reached dangerous levels [45]. Similar to the results of 2001, the RSIDI obtained could effectively describe the time and spatial occurrence patterns of the SPI and the WBDI, while the ADCI was analyzed to have delayed drought due to changes in SMSI. The ROC statistical analysis also confirmed that WBDI was highest and ADCI was relatively low in 2014 (SPI: 0.81, ADCI: 0.61, WBDI: 0.88). Through evaluation of past droughts, the RSIDI explained the three drought characteristics (meteorological, agricultural, and hydrological) well and confirmed the applicability of the integrated drought index through ROC analysis.

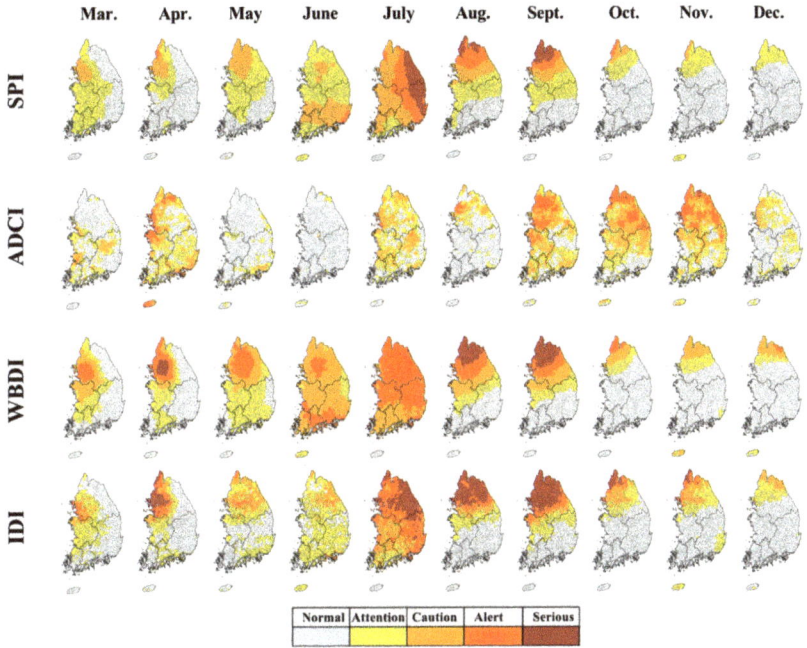

Figure 7. Spatial change of each drought index: SPI (standardized precipitation index), ADCI (agricultural dry condition index), WBDI (water budget-based drought index), and IDI (integrated drought index) for the 2014 drought.

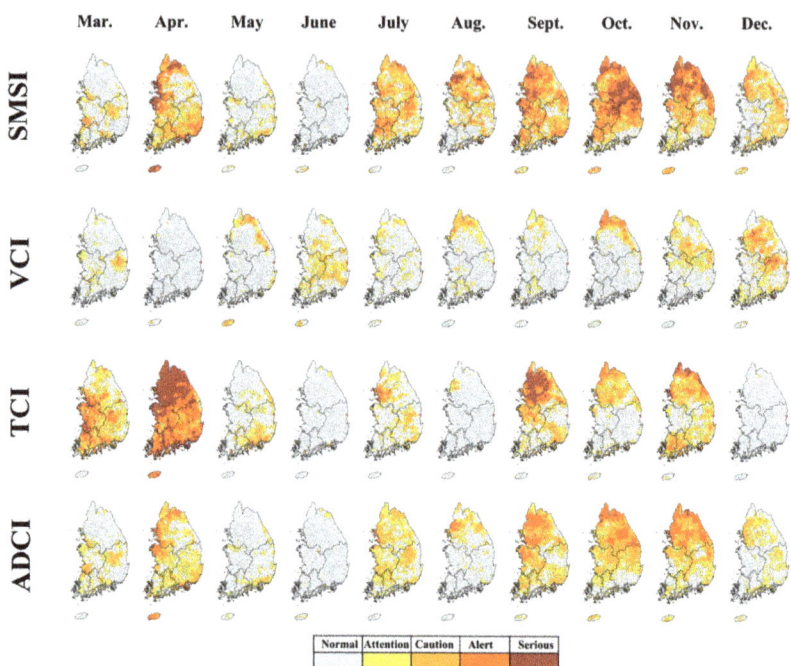

Figure 8. Spatial change of each drought index: SMSI, VCI, TCI, and ADCI for the 2014 drought.

3.2. Drought Transition Evaluation by Statistical Simulations

This study simulated the impact on the spatiotemporal distribution of different drought conditions, such as agricultural and hydrological perspectives, by altering the intended difference of $\delta_\mu(r)$ (Equation (11)) of the meteorological drought index (SPI) 1000 times to represent the changes in SPI. As $\delta_\mu(r)$ was intentionally changed, all the sub-basins experienced various changes in their ADCI and WBDI. Figure 9 shows that statistical simulations indicate that the changes in the state of agricultural (ADCI) and hydrological (WBDI) droughts correspond to the changes in meteorological drought (SPI). In addition, the results of the IBB simulation for up to three months of a lagged analysis were shown in Figure 10. Natural disasters, including droughts, are managed in four stages (Attention, Caution, Alert, and Serious) in Korea. In this study, the changes in other drought conditions were identified by intentionally changing the meteorological drought condition by using a IBB simulation.

First, when the meteorological drought (SPI) conditions were simulated from a stage of Normal to a stage of Attention (Figures 9a and 10a), the ADCI conditions, except for some parts of the Han River basin, showed a stage of Attention (96.6%, 820 out of 849 sub-basins). According to the results of the one-month delay, the ADCI status in parts of the Han and Nakdong River basins changed to Normal; however, 77.1% of the entire basin (45.8% of the two-month delay) was still in a stage of Attention. Over time, the ADCI conditions have shifted from a state of Attention to Normal in the southwestern part of the Han, Youngsang River, and Sumjin River basins. Even if the SPI conditions are simulated from the Normal stage to a stage of Attention, the results of the SPI are similar to those of time and space, and there are many sub-basins that are converted to the Normal stage over time. When the meteorological drought (SPI) conditions were simulated from the Normal stage to a stage of Caution (Figures 9b and 10b), excluding the parts of the Han River basin, more than 98.6% showed the ADCI to be in the Attention stage and 15.9% showed a stage of Caution. According to the results of the one-month delay, the ADCI in the northern parts of the Han River was still in a state of Caution; however, 88.5% of the entire basin was in the stage of Attention. Over time, the ADCI conditions shifted from a state of Attention to Normal in the southwestern parts of the Han, Youngsan, and Sumjin River basins. Although the mid-term drought (SPI6) was simulated from a Normal stage to a stage of Attention, changes in the spatiotemporal pattern of the ADCI were similar to the results obtained for SPI3, in which the scope of Attention conditions was reduced, and the number of sub-basins converted to "Normal" over time increased in the southern parts of the country.

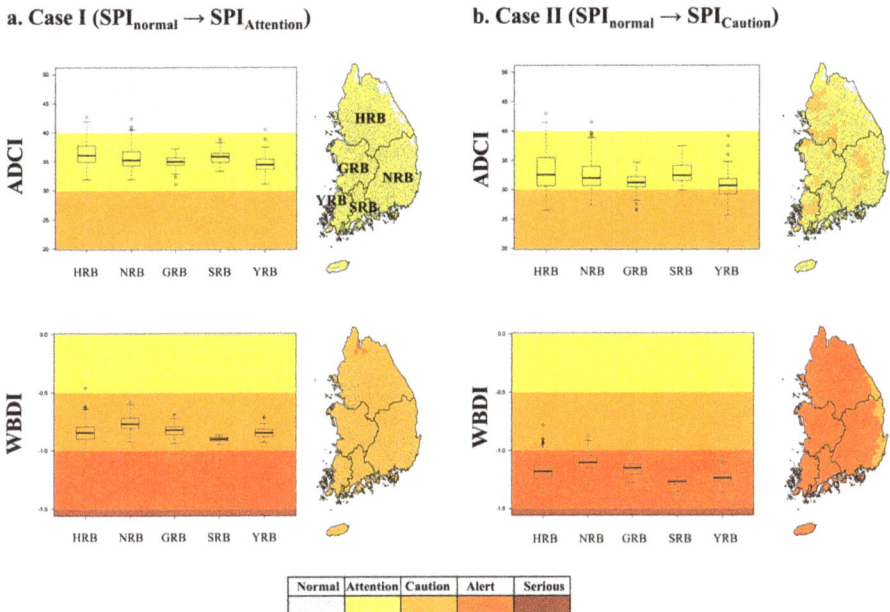

Figure 9. Drought condition changes in intentionally biased bootstrap (IBB) simulation. (**a**) Case I (SPI$_{normal}$ → SPI$_{Attention}$) and (**b**) Case II (SPI$_{normal}$ → SPI$_{Caution}$). The simulation results for the five basin areas are provided in the boxplot, and the results for the change in drought conditions are indicated in color in the map.

For WBDI, when the SPI drought conditions were simulated from Normal to a stage of Attention, the WBDI conditions, excluding the Han River basins, showed a stage of Alert (99.2%, 842 out of 849 sub-basins). As a result of the one-month delay, the WBDI status shifted from 75.7% of the total basin to a state of Attention, but 50.7% of the Han River basin and some areas of the Geum and Youngsan River basins were still in Caution levels. Over time, the WBDI conditions tended to shift from a state of Attention to Normal in the southwestern parts of the Han and Geum River basins. The SPI drought conditions were simulated from Normal to Caution; more than 93.8% of the total basins showed their WBDI in a stage of Caution. Over time, the WBDI status shifted to a Normal state around the southwestern Han and Geum River basins.

When the drought conditions of the RSIDI were simulated from Normal to Attention (Figure 11a), in all three drought indices (SPI, ADCI, and WBDI), the drought conditions shifted to a state of Attention, confirming that the RSIDI expressed the overall drought in space effectively. If drought conditions were simulated from Normal to Caution, the SPI results showed that 78.6% of the total basins were in the same state as drought conditions in the RSIDI. However, we inferred that the Han River basin was relatively insensitive due to its status of Attention. Compared to the drought conditions of the SPI, the drought conditions of the ADCI and WBDI were shown to be mitigated by one level in the Han River and some areas of the Nakdong River, and the spatial conditions changed in the two drought indices (ADCI and WBDI) were similar.

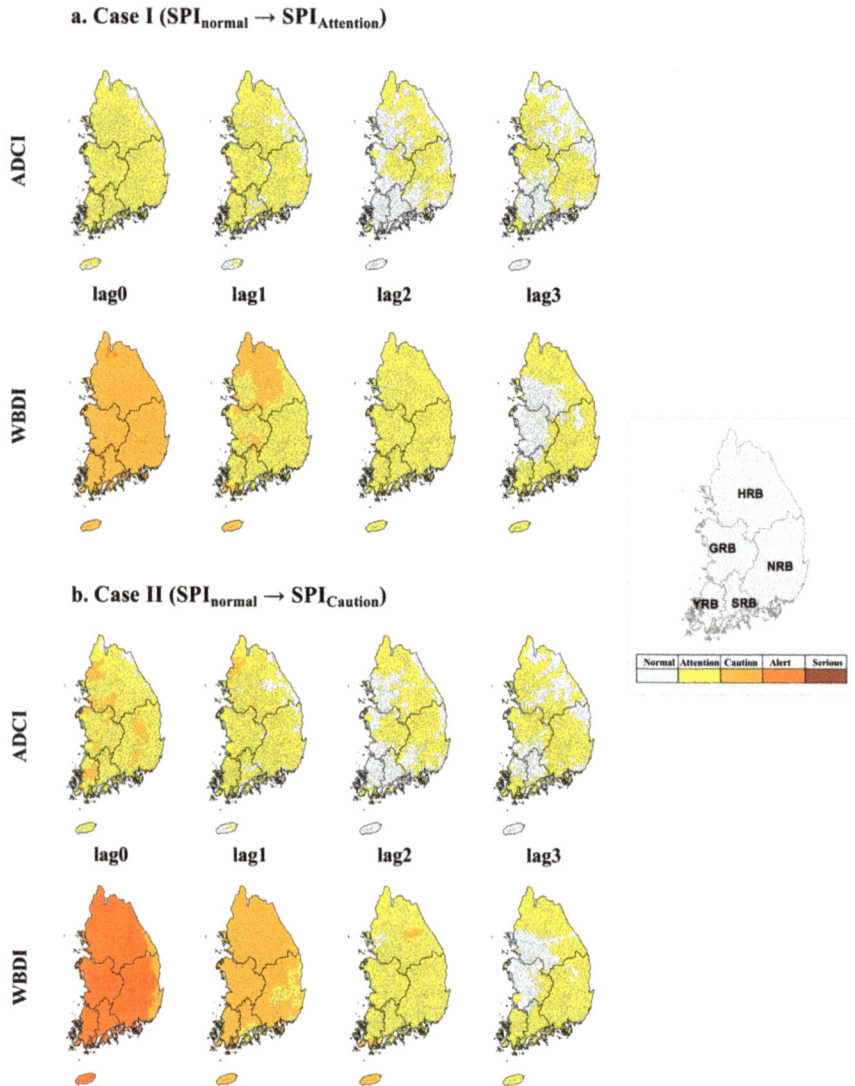

Figure 10. Drought condition changes in IBB simulation of ADCI and WBDI. (**a**) Case I (SPI$_{normal}$ → SPI$_{Attention}$) and (**b**) Case II (SPI$_{normal}$ → SPI$_{Caution}$). The simulation results for the five basin areas for up to three months are colored in the map.

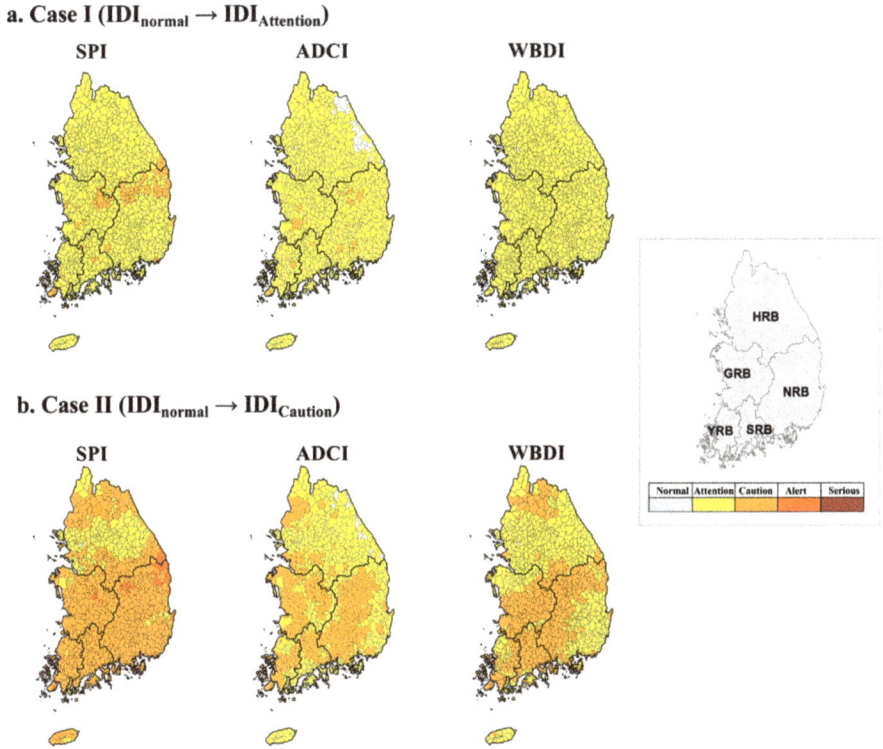

Figure 11. Drought condition changes in IBB simulation. (**a**) Case I (IDI$_{normal}$ → IDI$_{Attention}$) and (**b**) Case II (IDI$_{normal}$ → IDI$_{Caution}$). The simulation results for the five basin areas for up to three months are colored in the map.

4. Discussion and Conclusions

As climate change accelerates due to global warming, changes in hydrological cycles occur significantly, and water use and prediction of water resources may become difficult. In particular, in Korea, chronic drought has occurred continuously since the 1990s during the transition from winter to spring [46]. To cope with these droughts, technologies to identify and predict the magnitude of spatiotemporal droughts are required. Drought monitoring using satellite data will be essential to secure spatial resolution for accurate and spatial droughts when ground-based hydrometeorological data are not available, as well as monitoring the different characteristics of meteorological, agricultural, and hydrological droughts.

Korea's drought-related affairs are mainly handled by the Korea Meteorological Administration (KMA), Ministry of Agriculture, Food and Rural Affairs (MAFRA), Ministry of Environment (MOE), Ministry of Land, Infrastructure, and Transport (MOLIT), and the Ministry of Public Safety and Security (MPSS) [47]. The KMA diagnoses precipitation and drought in drought areas by assessing the SPI and PDSI and provides this information to the local governments. The MAFRA analyzes agricultural water through the soil moisture index (SMI), reservoir drought index (RDI), and integrated agricultural drought index. The MOLIT monitors dam water; the MOE monitors emergency water resources and water quality according to the drought stage and implements the appropriate countermeasures. The MPSS oversees the drought situation when it becomes extreme. However, some point out that the current drought measurement indices of different agencies are different in drought management, causing confusion and making it difficult to respond to drought proactively. Different ministries have different standards for determining the degree of drought. Additionally, it is not sufficient for a single drought indicator to characterize all the complex drought evolution processes [22]. The development and application of an integrated drought index are necessary to take into account the complex factors related to water use, such as the meteorological, agricultural, and hydrological perspectives. Thus, this study proposed an RS-based integrated drought index that was extracted from 849 sub-basins in Korea's five major river basins using multi-sensor collaborative approaches and multivariate dimensional reduction models, calculated through monthly satellite data. Droughts in 2001 and 2014, representative years of severe drought since the 2000s, were evaluated using the integrated drought index, and statistical simulations were used to diagnose the sensitivity and transition of drought. The BPCA-based integrated drought index proposed in this study was analyzed to reflect the timing, severity, and evolutionary pattern of meteorological, agricultural and hydrological droughts, enabling comprehensive delivery of drought information. Although the results relied on limited observations, it is expected that drought hotspot analyses and statistical simulations using IBB and BPCA-based RSIDI will identify the drought characteristics of the sub-basin, thereby promoting their use in preemptive drought response through drought prediction and early warning.

Drought monitoring and accurate drought forecasting are still the main challenges in a relatively changing environment that has a long lead-time and natural and artificial factors. Therefore, future works to improve drought monitoring and prediction require further research, such as high-quality data assimilation, improving model development through major processes related to droughts, selecting or predicting optimal ensembles, and hybrid drought forecasting.

Author Contributions: Conceptualization, Resources, Formal analysis, Writing—original draft, J.-S.K. and S.-Y.P.; Conceptualization, Methodology, Writing—review & editing, J.-H.L.; Writing—review & editing, J.C., S.C. and T.-W.K. All authors have read and agreed to the published version of the manuscript.

Funding: This research was supported by the Korea Environment Industry & Technology Institute (KEITI) through the Water Management Research Program, which is funded by the Korea Ministry of Environment (MOE) (Grant No. 79616). We also appreciate the support of the State Key Laboratory of Water Resources and Hydropower Engineering Science, Wuhan University.

Conflicts of Interest: The authors declare no conflict of interest.

References

1. Zarafshani, K.; Sharafi, L.; Azadi, H.; Van Passel, S. Vulnerability Assessment Models to Drought: Toward a Conceptual Framework. *Sustainability* **2016**, *8*, 588. [CrossRef]
2. Kim, J.S.; Seo, G.S.; Jang, H.W.; Lee, J.H. Correlation analysis between Korean spring drought and large-scale teleconnection patterns for drought forecasting. *KSCE J. Civ. Eng.* **2017**, *21*, 458–466. [CrossRef]
3. Yue, Y.; Shen, S.; Wang, Q. Trend and variability in droughts in Northeast China based on the reconnaissance drought index. *Water* **2018**, *10*, 318. [CrossRef]
4. Qaiser, G.; Tariq, S.; Adnan, S.; Latif, M. Evaluation of a composite drought index to identify seasonal drought and its associated atmospheric dynamics in Northern Punjab, Pakistan. *J. Arid Environ.* **2021**, *185*, 104332. [CrossRef]
5. Cong, D.; Zhao, S.; Chen, C.; Duan, Z. Characterization of droughts during 2001–2014 based on remote sensing: A case study of Northeast China. *Ecol. Inf.* **2017**, *39*, 56–67. [CrossRef]
6. Park, S.Y.; Sur, C.; Kim, J.S.; Lee, J.H. Evaluation of multi-sensor satellite data for monitoring different drought impacts. *Stoch. Environ. Res. Risk Assess.* **2018**, *32*, 2551–2563. [CrossRef]
7. Sur, C.; Park, S.Y.; Kim, T.W.; Lee, J.H. Remote sensing-based agricultural drought monitoring using hydrometeorological variables. *KSCE J. Civ. Eng.* **2019**, *23*, 5244–5256. [CrossRef]
8. Abuzar, M.K.; Shafiq, M.; Mahmood, S.A.; Irfan, M.; Khalil, T.; Khubaib, N. Drought risk assessment in the khushab region of Pakistan using satellite remote sensing and geospatial methods. *Int. J. Econ. Environ. Geol.* **2019**, *10*, 48–56. [CrossRef]
9. Zhong, R.; Chen, X.; Lai, C.; Wang, Z.; Lian, Y.; Yu, H.; Wu, X. Drought monitoring utility of satellite-based precipitation products across mainland China. *J. Hydrol.* **2019**, *568*, 343–359. [CrossRef]
10. Carlson, T.N.; Gillies, R.R.; Perry, E.M. A method to make use of thermal infrared temperatureand NDVI measurements to infer surface soil water content and fractional vegetation cover. *Remote Sens. Rev.* **1994**, *9*, 161–173. [CrossRef]

11. Tadesse, T.; Demisse, G.B.; Zaitchik, B.; Dinku, T. Satellite-based hybrid drought monitoring tool for prediction of vegetation condition in Eastern Africa: A case study for Ethiopia. *Water Resour. Res.* **2014**, *50*, 2176–2190. [CrossRef]
12. Enenkel, M.; Steiner, C.; Mistelbauer, T.; Dorigo, W.; Wagner, W.; See, L. A combined satellite-derived drought indicator to support humanitarian aid organizations. *Rem. Sens.* **2016**, *8*, 340. [CrossRef]
13. Wang, J.; Xu, X.; Ding, S.; Zeng, J.; Spurr, R.; Liu, X.; Chance, K.; Mishchenko, M. A numerical testbed for remote sensing of aerosols, and its demonstration for evaluating retrieval synergy from a geostationary satellite constellation of GEO-CAPE and GOES-R. *J. Quant. Spectrosc. Radiat. Transfer.* **2014**, *146*, 510–528. [CrossRef]
14. Wilhite, D.A. *Drought Monitoring and Early Warning: Concepts, Progress and Future Challenges*; World Meteorological Organization: Geneva, Switzerland, 2006; p. 1006.
15. Huang, S.; Huang, Q.; Leng, G.; Liu, S. A nonparametric multivariate standardized drought index for characterizing socioeconomic drought: A case study in the Heihe River Basin. *J. Hydrol.* **2016**, *542*, 875–883. [CrossRef]
16. Wu, Z.; Mao, Y.; Li, X.; Lu, G.; Lin, Q.; Xu, H. Exploring spatiotemporal relationships among meteorological, agricultural, and hydrological droughts in Southwest China. *Stoch Environ. Res. Risk Assess.* **2016**, *30*, 1033–1044. [CrossRef]
17. Sur, C.; Park, S.Y.; Kim, J.S.; Lee, J.H. Prognostic and diagnostic assessment of hydrological drought using water and energy budget-based indices. *J. Hydrol.* **2020**, *591*, 125549. [CrossRef]
18. Guo, Y.; Huang, S.; Huang, Q.; Wang, H.; Fang, W.; Yang, Y.; Wang, L. Assessing socioeconomic drought based on an improved multivariate standardized reliability and resilience index. *J. Hydrol.* **2019**, *568*, 904–918. [CrossRef]
19. McKee, T.B.; Doesken, N.J.; Kleist, J. The relationship of drought frequency and duration to time scales. In Proceedings of the 8th Conference on Applied Climatology, Anaheim, CA, USA, 17–22 January 1993; Volume 17, pp. 179–183.
20. Shukla, S.; Wood, A.W. Use of a standardized runoff index for characterizing hydrologic drought. *Geophys. Res. Lett.* **2008**, *35*, 1100. [CrossRef]
21. Hayes, M.; Svoboda, M.; Wall, N.; Widhalm, M. The Lincoln declaration on drought indices: Universal meteorological drought index recommended. *Bull. Am. Meteorol. Soc.* **2011**, *92*, 485–488. [CrossRef]
22. Hao, Z.; AghaKouchak, A. A nonparametric multivariate multi-index drought monitoring framework. *J. Hydrometeor.* **2014**, *15*, 89–101. [CrossRef]
23. Cunha, A.P.M.; Alvalá, R.C.; Nobre, C.A.; Carvalho, M. A Monitoring vegetative drought dynamics in the Brazilian semiarid region. *Agric. Meteorol.* **2015**, *2014*, 494–505. [CrossRef]
24. Sur, C.; Hur, J.; Kim, K.; Choi, W.; Choi, M. An evaluation of satellite-based drought indices on a regional scale. *Int. J. Remote Sens.* **2015**, *36*, 5593–5612. [CrossRef]
25. Li, J.; Zhou, S.; Hu, R. Hydrological drought class transition using SPI and SRI time series by loglinear regression. *Water Resour. Manage.* **2016**, *30*, 669–684. [CrossRef]
26. Palmer, W. *Meteorological Drought, Weather Bureau Research Paper 45*; U.S. Weather Bureau: Washington, DC, USA, 1965; p. 58.
27. Zhang, A.; Jia, G. Monitoring meteorological drought in semiarid regions using multi-sensor microwave remote sensing data. *Remote Sens. Environ.* **2013**, *134*, 12–23. [CrossRef]
28. Zhang, X.; Chen, N.; Li, J.; Chen, Z.; Niyogi, D. Multi-sensor integrated framework and index for agricultural drought monitoring. *Remote Sens. Environ.* **2017**, *188*, 141–163. [CrossRef]
29. Kogan, F.N. Application of vegetation index and brightness temperature for drought detection. *Adv. Space Res.* **1995**, *15*, 91–100. [CrossRef]
30. Kogan, F.N. Operational space technology for global vegetation assessment. *Bull. Am. Meteorol. Soc.* **2001**, *82*, 1949–1964. [CrossRef]
31. Oki, T.; Kanae, S. Global hydrological cycles and world water resources. *Science* **2006**, *313*, 1068–1072. [CrossRef]
32. Huffman, G.J.; Adler, R.F.; Bolvin, D.T.; Gu, G.; Nelkin, E.J.; Bowman, K.P.; Hong, Y.; Stocker, E.F.; Wolff, D.B. The TRMM Multi-Satellite Precipitation Analysis: Quasi-Global, Multi-Year, Combined-Sensor Precipitation Estimates at Fine Scale. *J. Hydrometeor.* **2007**, *8*, 38–55. [CrossRef]
33. McKee, T.B. Drought monitoring with multiple time scales. In Proceedings of the 9th Conference, Applied Climatology, Dallas, TX, USA, 15–20 January 1995; pp. 233–236.
34. Hao, Z.; AghaKouchak, A. Multivariate Standardized Drought Index: A prametric multi-index model. *Adv. Water Resour.* **2013**, *57*, 12–18. [CrossRef]
35. Ma, Y.Y.; Zhu, L.P. A Review on Dimension Reduction. *Int. Stat. Rev.* **2012**, *81*, 134–150. [CrossRef] [PubMed]
36. Jolliffe, I.T. *Principal Component Analysis*, 2nd ed.; Springer Science Business Media: Berlin, Germany, 2002.
37. Oba, S.; Sato, M.A.; Takemasa, I.; Monden, M.; Matsubara, K.I.; Ishii, S. A Bayesian missing value estimation method for gene expression profile data. *Bioinformatics* **2003**, *19*, 2088–2096. [CrossRef] [PubMed]
38. Lai, W.Y.; Kuok, K.K. A Study on Bayesian Principal Component Analysis for Addressing Missing Rainfall Data. *Water Resour. Manag.* **2019**, *33*, 2615–2628. [CrossRef]
39. Lee, T. Climate change inspector with intentionally biased bootstrapping (CCIIBB ver. 1.0)–methodology development. *Geosci. Model. Dev.* **2017**, *10*, 525–536. [CrossRef]
40. Bouveyron, C.; Latouche, P.; Mattei, P.A. Exact dimensinality selection for Bayesian PCA. *Scand. J. Statist.* **2020**, *47*, 196–211. [CrossRef]

41. Heng, C.; Lee, T.; Kim, J.-S.; Xiong, L. Influence analysis of central and Eastern Pacific El Niños to seasonal rainfall patterns over China using the intentional statistical simulations. *Atmos. Res.* **2020**, *233*, 104706. [CrossRef]
42. Zelinka, I. SOMA—self-organizing migrating algorithm. In *New Optimization Techniques in Engineering*; Springer: Berlin/Heidelberg, Germany, 2004; pp. 167–217.
43. Baek, S.G.; Jang, H.W.; Kim, J.S.; Lee, J.H. Agricultural drought monitoring using the satellite-based vegetation index, Korea Water Resources Association. *J. Korea Water Resour. Assoc.* **2016**, *49*, 305–314. (In Korean) [CrossRef]
44. Lee, J.H.; Jang, H.W. Comparison on Characteristics and Historical Drought Events of summer drought in 2014. Korea Disaster Prevention Association. *J. Disaster Prev.* **2014**, *16*, 46–56. (In Korean)
45. Ministry of Land, Infrastructure and Transport (MLIT). *2015 Drought Impact Investigation Report*; Korea Ministry of Land, Infrastructure and Transport (MLIT): Sejong City, Korea, 2015. (In Korean)
46. Bae, H.; Ji, H.; Lim, Y. Characteristics of drought propagation in South Korea: Relationship between meteorological, agricultural, and hydrological droughts. *Nat. Hazards.* **2019**, *99*, 1–16. [CrossRef]
47. Hong, I.P.; Lee, J.H.; Cho, H.S. National drought management framework for drought preparedness in Korea (lessons from the 2014–2015 drought). *Water Policy* **2016**, *18*, 89–106. [CrossRef]

Article

Statistical Prediction of Typhoon-Induced Rainfall over China Using Historical Rainfall, Tracks, and Intensity of Typhoon in the Western North Pacific

Jong-Suk Kim [1,*], Anxiang Chen [1], Junghwan Lee [2], Il-Ju Moon [3] and Young-Il Moon [2]

[1] State Key Laboratory of Water Resources and Hydropower Engineering Science, Wuhan University, Wuhan 430072, China; chenanxiang9@gmail.com
[2] Urban Flood Research Institute, University of Seoul, Seoul 02504, Korea; jhlee88@uos.ac.kr (J.L.); ymoon@uos.ac.kr (Y.-I.M.)
[3] Typhoon Research Center/Graduate School of Interdisciplinary Program in Marine Meteorology, Jeju National University, Jeju 63243, Korea; ijmoon@jejunu.ac.kr
* Correspondence: jongsuk@whu.edu.cn

Received: 19 November 2020; Accepted: 11 December 2020; Published: 17 December 2020

Abstract: Typhoons or mature tropical cyclones (TCs) can affect inland areas of up to hundreds of kilometers with heavy rains and strong winds, along with landslides causing numerous casualties and property damage due to concentrated precipitation over short time periods. To reduce these damages, it is necessary to accurately predict the rainfall induced by TCs in the western North Pacific Region. However, despite dramatic advances in observation and numerical modeling, the accuracy of prediction of typhoon-induced rainfall and spatial distribution remains limited. The present study offers a statistical approach to predicting the accumulated rainfall associated with typhoons based on a historical storm track and intensity data along with observed rainfall data for 55 typhoons affecting the southeastern coastal areas of China from 1961 to 2017. This approach is shown to provide an average root mean square error of 51.2 mm across 75 meteorological stations in the southeast coastal area of China (ranging from 15.8 to 87.3 mm). Moreover, the error is less than 70 mm for most stations, and significantly lower in the three verification cases, thus demonstrating the feasibility of this approach. Furthermore, the use of fuzzy C-means clustering, ensemble averaging, and corrections to typhoon intensities, can provide more accurate rainfall predictions from the method applied herein, thus allowing for improvements to disaster preparedness and emergency response.

Keywords: typhoon-induced rainfall; prediction; statistical model; fuzzy C-means clustering; China

1. Introduction

Coastal areas with high population densities and rapid growth and urbanization have relatively vulnerable structures to coastal flooding, such as the sea-level rise and storm surge due to climatic extremes [1,2]. The losses caused by these disasters have also continued to increase in recent years. The western North Pacific (WNP) is one of the oceanic regions most prone to typhoons [3–9]. Since China is located on the west coast of the WNP, it is greatly affected by typhoons, particularly along the east coast [10]. The strong winds, heavy precipitation, and storm surge of typhoons pose serious threats to China's social economy and national personal safety. For example, the super typhoon "Mangkhut" affected many provinces and regions over South China in September 2018. The number of people affected was close to 3 million, with ~1200 houses damaged and ~174.4 thousand hectares of crops being affected. The direct economic loss exceeded CNY 5.2 billion (USD 77.5 million) [11].

Failure to properly manage water resources due to incorrect rainfall forecasts during the typhoon season can lead to serious flooding or water shortage, regardless of how well forecast and water management was carried out before the typhoon [10,12]. In recent years, however, the development of satellite observations and mathematical modeling, along with integration and data assimilation techniques using various observational datasets, typhoon tracking and intensity prediction have continuously improved [13–20]. Nevertheless, typhoon-induced rainfall prediction remains very difficult and less accurate than typhoon track prediction [21–29]. For example, Li et al. [26] established a non-parametric statistical method using numerical models and typhoon intensity predictions to estimate the maximum daily rainfall and three-day cumulative rainfall amounts. Previously, Ebert et al. [27] noted that a satellite-based tracking of tropical rain could improve the short-term prediction of typhoon-induced heavy rainfall. More recently, Kim et al. [28] hypothesized that typhoons with similar tracks have similar rainfall patterns, and demonstrated the use of tracks, intensities, and precipitation data for 91 typhoons affecting the Korean Peninsula over the course of several decades to establish a statistical model for forecasting typhoon-induced rainfall over that region.

Although typhoon-induced rainfall prediction models are constantly being improved, the rainfall conditions related to typhoons differ from region to region and most of the aforementioned methods were developed according to one or other specific regions [26–30]. While the establishment of a typhoon-induced rainfall prediction model requires accurate track and intensity forecasts; however, complex physical processes such as the interaction between typhoon and land also need to be considered. These factors may cause rapid changes in precipitation during the passage of typhoons [21,22]. Therefore, typhoon-induced rainfall prediction is particularly challenging work.

The purpose of the present study is to establish a new statistical prediction model based on the principle of track similarity, using fuzzy C-means clustering, intensity correction, and other methods to optimize typhoon-induced accumulated rainfall (TAR) forecasts over China. The following section introduces the data used to develop the prediction model and describes how the TAR of each typhoon in the western North Pacific in recent decades is determined. Then, in Section 3, typhoons with tracks similar to that of the target typhoon are selected. In addition, TAR correction is conducted based on typhoon intensity, and the optimal number of similar-track typhoons is selected for ensemble averaging. After substituting the previous typhoon data, the results of the prediction model are given. Finally, Section 4 provides a summary and conclusions, including a discussion of the advantages of this method as well as the limitations that can be improved in future work.

2. Data and Methods

2.1. Data

To establish the TAR prediction model, the daily rainfall data without any gaps between 1961 and 2017 from 537 meteorological stations in China (Figure 1a; http://data.cma.cn) were used, along with best-track data for a total of 1536 typhoons in the WNP were used during the period 1961–2017 (Figure 1b). Typhoon intensity correction was performed and the effects of ensemble averaging and typhoon similarity levels were analyzed using primarily the 55 tropical cyclone (TC) datasets affecting 75 meteorological stations in the southeast coastal area of China listed in Table 1. The 6-hourly location and intensity data for the typhoons, including the specific date, time, longitude, latitude, maximum wind speed, and typhoon number, were obtained from the Regional Specialized Meteorological Center (RSMC)—Tokyo.

Due to the proximity of typhoons to mid-latitude regions, typhoons will transition into tropical storms under the impacts of landfall, cold air mixing, and other factors, leading to a rapid weakening of their intensities. Nevertheless, the impact of the associated rainfall will impact large areas and generate disasters such as debris flows and floods that may cause losses of life and property. Therefore, in order to better estimate the rainfall that a typhoon can cause, the present study includes the period after each typhoon turns into a tropical storm.

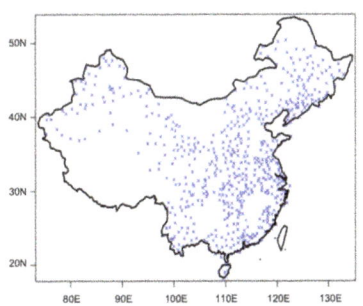
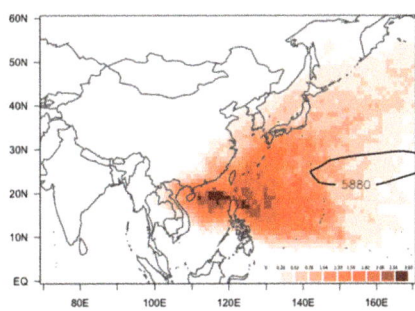

Figure 1. Weather stations and typhoon track data used in the present study: (**a**) the locations of the meteorological stations ($n = 537$); and (**b**) the long-term average of the tropical cyclone (TC) track density in the western North Pacific (WNP) region during the period 1961–2017. The solid line in (**b**) indicates the location of the WNP subtropical high represented by 5880 gpm during the study period.

Table 1. The 55 typhoons used in the present study.

No	Name	ID	Maximum Wind (kt)	No	Name	ID	Maximum Wind (kt)
1	FREDA	7713	55	29	MORAKOT	0309	45
2	ROSE	7804	45	30	VAMCO	0311	35
3	DELLA	7812	45	31	DUJUAN	0313	80
4	GORDON	7908	55	32	KOMPASU	0409	45
5	PERCY	8014	100	33	RANANIM	0413	80
6	MAURY	8108	50	34	AERE	0417	80
7	IRVING	8217	85	35	SANVU	0510	50
8	WAYNE	8304	110	36	BILIS	0604	60
9	WYNNE	8402	55	37	BOPHA	0609	55
10	VAL	8517	45	38	WUTIP	0707	35
11	ELLEN	8620	70	39	NURI	0812	75
12	GERALD	8714	80	40	HAGUPIT	0814	90
13	NONAME	8803	35	41	MOLAVE	0906	65
14	FAYE	8907	55	42	FANAPI	1011	95
15	GORDON	8908	100	43	NANMADOL	1111	100
16	DOT	9017	75	44	SAOLA	1209	70
17	AMY	9107	95	45	CIMARON	1308	40
18	BRRENDAN	9108	60	46	UTOR	1311	105
19	GARY	9207	55	47	TRAMI	1312	60
20	BECKY	9316	55	48	USAGI	1319	110
21	TIM	9405	95	49	KALMAEGI	1415	75
22	CAITLIN	9412	45	50	LINFA	1510	50
23	HELEN	9505	50	51	NIDA	1604	60
24	GLORIA	9608	65	52	CHANTHU	1617	85
25	SALLY	9616	85	53	PAKHAR	1714	55
26	OTTO	9802	65	54	GUCHOL	1717	35
27	MAGGIE	9903	75	55	KHANUN	1720	75
28	TRAMI	0105	40				

2.2. Calculation of Typhoon-Induced Accumulated Rainfall (TAR)

The first step in establishing the TAR prediction model is to calculate the TAR for each local station. The specific calculation process is as follows:

(1). Determine whether the rainfall for a specific location is caused by the typhoon. Only when the distance between a typhoon and a specific location is less than a certain value can the typhoon be considered to have an impact on that location's rainfall. The selection criterion used in the present study is that the distance between the typhoon and the meteorological station must be less than or equal to 500 km. [9,31–33]
(2). As the typhoon will have a continuous impact on the rainfall in a specific area, the rainfall on the day in which the region is affected is considered, along with that of the day before and the day after, as being caused by the typhoon. That is, the total duration of rainfall caused by a typhoon in a specific location is represented by the time period of the typhoon entering and leaving a 500 km range of the area within a time window ± 1 day of its landfall.
(3). The TAR values for each typhoon at each station are obtained by adding up the daily rainfall in the previously determined period.

It was noted that a substantial error would arise if coexisting typhoons were used to establish a TAR prediction model, which would result in an inaccurate model forecast. To prevent this problem, typhoons of this type were discarded during the prediction model establishment process.

2.3. Selection of Typhoons Using the Fuzzy C-Means Clustering Algorithm

In the present study, the fuzzy C-means clustering (FCM) algorithm was used to select typhoons with similar tracks. This is a partitioning algorithm in which objects with the greatest similarities are grouped into the same cluster and objects with few similarities into separate clusters. The FCM was proposed by Bezdek [34] as an improvement on the hard C-means clustering method and enables an estimate of the degree to which each data point belongs to a certain cluster, i.e., the degree of membership. In detail, the FCM divides n vectors X_i (i = 1, 2,..., n) into a number (c) of fuzzy groups and identifies the clustering center of each group so that the value function of the dissimilarity index is minimized. A fuzzy division is then used to assign a degree of membership between 0 and 1 and examine how well each data point belongs to each group. According to the FCM, the membership matrix U assigns the values of the elements between 0 and 1, while the constraints of the normalization dictate that the total membership of the dataset must always be equal to unity, as indicated by Equation (1):

$$\sum_{i=1}^{c} u_{ij} = 1, \quad \forall j = 1, \cdots, n \tag{1}$$

Then, the value function (or objective function) of the FCM is given by Equation (2):

$$J(U, c_1, \cdots, c_c) = \sum_{i=1}^{c} J_i = \sum_{i=1}^{c} \sum_{j}^{n} u_{ij}^m d_{ij}^2 \tag{2}$$

where u_{ij} is between 0 and 1, c_i is the clustering center of the fuzzy group i, and $d_{ij} = \|c_i - x_i\|$ is the Euclidean distance between the i-th clustering center and the j-th data point.

In the process of clustering typhoons using the FCM method, the membership coefficient W_{ik} is calculated. This indicates the probability, X_i, that each typhoon belongs to the target typhoon group C_k [28,35]. The value of W_{ik} is determined by the partial derivative of the sum of squared errors (SSE) according to Equations (3) and (4):

$$SSE = \sum_{k=1}^{K} \sum_{i=1}^{n} W_{ik}{}^p d(x_i, c_k)^2 \tag{3}$$

$$W_{ik} = \frac{\{\frac{1}{d(x_i,c_k)^2}\}^{\frac{1}{p-1}}}{\sum_{K=1}^{K}\{\frac{1}{d(x_i,c_k)^2}\}^{\frac{1}{p-1}}} \tag{4}$$

where $d(x_i,c_k)^2$ is the distance between each typhoon track and the target typhoon track.

When using the FCM method to cluster all the typhoon tracks, these must first be divided into lines with the same number of location points. In the present study, all typhoons were uniformly interpolated according to the typhoon with the largest number of location points in its track data. In addition, the FCM membership coefficient was used as a criterion for screening typhoons that were similar to the target typhoon: the larger the coefficient value, the higher the typhoon similarity. For example, the eight typhoons with the greatest similarity to typhoons Usagi (#1319) and Nesat (#1117) according to the FCM method are indicated in Figure 2. Typhoon Usagi (#1319) made landfall on the coast of Fujian Province in southern China in 2013 and affected the surrounding areas of Taiwan and southern provinces of China (Figure 2a), whereas typhoon Nesat (#1117) passed through Hainan Province and Qiongzhou Strait in 2011 and then caused serious damage to the surrounding areas, including Hainan, Guangdong, and other provinces (Figure 2b). The results in Figure 2 indicate that the tracks of typhoons Nuri (#0812) in 2008 and Sharon (#9404) in 1994 are the most similar to those of Usagi (#1319) and Nesat (#1117), respectively.

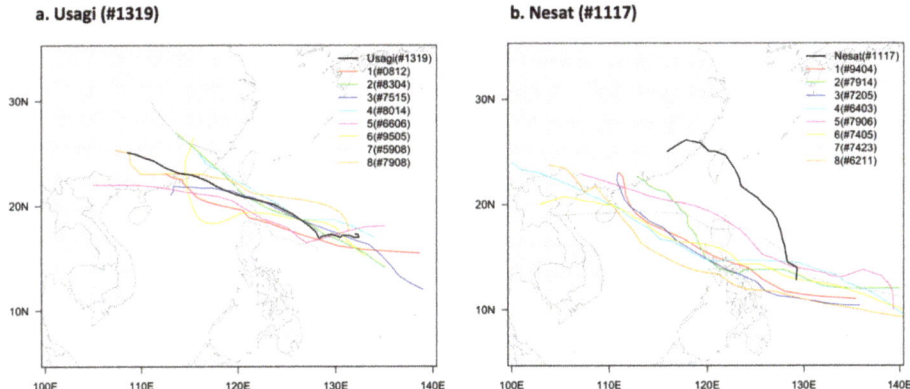

Figure 2. Top eight typhoon tracks most similar to those of: (**a**) Usagi (#1319) and (**b**) typhoon Nesat (#1117). The identification number and similarity level of the selected typhoons are indicated in the key.

3. Results

3.1. Correcting the TAR Using Typhoon Intensity Information

Since it is impossible for different typhoons to have exactly the same intensity and structure, every typhoon is unique. Therefore, it is not theoretically possible to accurately predict the amount of rainfall caused by a typhoon based only on the track of one typhoon only. In other words, even when two typhoons have exactly the same tracks, differences in their intensities will result in different rainfall amounts, with higher intensity typhoons usually resulting in more rainfall [36]. Therefore, a typhoon wind intensity correction (TWIC) was used in the present study to further reduce the error in the TAR prediction model. The effects of the TWIC and ensemble averaging were first assessed using the training datasets of 55 TCs and then verified for model performance later in Section 3.3.

The eastern and southern coastal areas of China were selected as target areas for prediction during the training of this model because these are the areas that are most frequently affected by typhoons, whereas the inland areas of China are rarely affected. In the process of TAR correction based on TC wind speed, data from typhoons affecting 75 weather stations along the southeastern coast of China

(Pearl River Basin and Southeast River Basin) were used. Typhoons that occur simultaneously in the same region were not used for this process, as it is difficult to obtain their individual TAR periods and rainfall amounts accurately.

After processing the data, the 55 most representative typhoons with high data accuracy and their corresponding similar-track typhoons were finally selected. The TC wind speed and average rainfall values during the passage of these typhoons were then calculated from the data obtained from 75 stations in the southeast coastal area of China. Using these data, the linear regression equation relating the TC wind speed of the 75 weather stations and the average TAR during typhoon passage was obtained (Figure 3) and the best fit was given by Equation (5):

$$\bar{P}_{TAR} = 0.654V + 10.891 \tag{5}$$

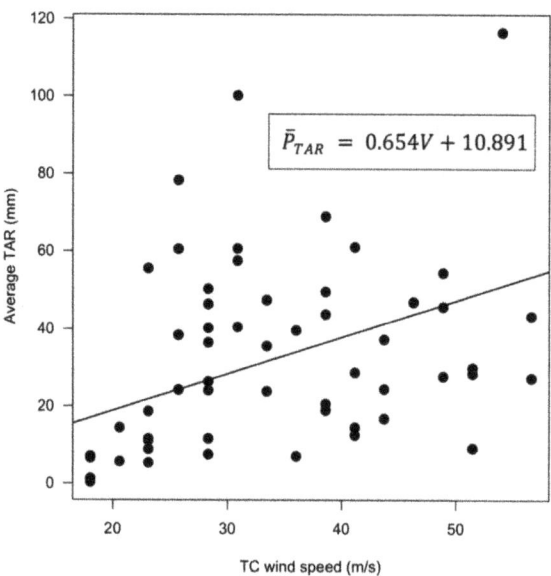

Figure 3. Linear relationship between the TC wind speed (V, m/s) and the average typhoon-induced accumulated rainfall (TAR) ($\bar{P}_{TAR}$, mm). The average TAR and TC wind speed were obtained using the most similar typhoons from 55 storms and 75 stations.

The equation shows that there is a positive correlation between the TC wind speed (V, m/s) and the average TAR ($\bar{P}_{TAR}$, mm). This shows a significant relationship ($p < 0.05$) between the TC wind speed and the average TAR ($R^2 = 0.654 \pm 0.291$). During the training process for the TAR prediction model, this linear equation was adopted to apply an intensity correction to all typhoons with similar tracks.

3.2. Effects of Track Similarity, Ensemble Averaging, and Intensity Correction on the TAR Predictions

The similarity level of the typhoon track, the number of ensemble averages, and whether the typhoon intensity is corrected may have an impact on the TAR prediction. To examine the influence of the typhoon track similarity level, the accuracy of the prediction result is judged by the root mean square error (RMSE), where a smaller error indicates a more accurate result. The results presented in Figure 4 (black line) show that the use of a single typhoon with the most similar track to predict the TAR values of the target typhoon in the target areas from 1961 to 2017 gives an average RMSE of 62.2 mm. However, if the typhoon with the second-best track similarity is used alone for the prediction, the RMSE is slightly decreased to 60.8 mm, while using only the typhoon rainfall data with the third-best track similarity decreases the average RMSE to 58.7 mm. Thereafter, the average RMSE continues to decrease as the similarity of the selected typhoon increases. In general, the prediction error decreases with the use of individual typhoons with increasing track similarity levels, but the use of only a single typhoon in the TAR prediction process may nevertheless result in an unsatisfactory error reduction even if its track is very similar to that of the target typhoon.

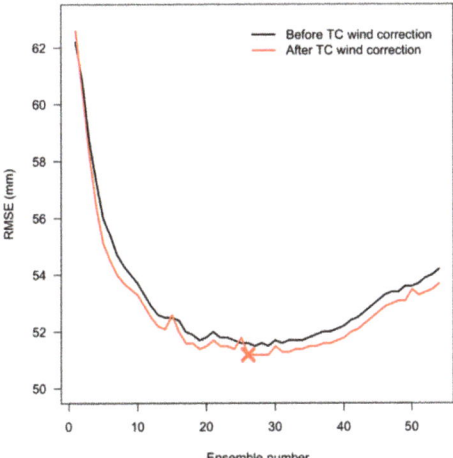

Figure 4. A comparison of the change in the RMSE of the prediction obtained with the increasing number of ensemble typhoons used before (black line) and after (red line) TC wind intensity correction. "X" represents the optimal ensemble number after TC wind intensity correction.

To further reduce the prediction errors, ensemble averaging (EA) was then considered [21,37]. To detect the influence of EA on the TAR prediction result, the number of high track-similarity typhoons used in the prediction at each station was increased step-by-step to form an ensemble, then their average TAR values were calculated and compared with the observed values. The results in Figure 4 (black line) indicate that as the number of typhoons in the ensemble increases, the RMSE initially decreases to a minimum of 51.5 mm with an ensemble of the 27 most similar typhoons, and gradually increases thereafter.

Then, to study the influence of typhoon wind intensity correction upon TAR prediction, the TAR obtained after TWIC was calculated using the EA method, and the results were compared with those obtained without TWIC in Figure 4. Here, the red line indicates a decrease of 0.5–0.9 mm in the average RMSE after the TWIC. In other words, the TWIC helps reduce the error in TAR predictions. In addition, the above results indicate an optimal ensemble number of 26 when using the EA method to predict the TAR.

Based on the results of the above analysis, the operational process of the statistical TAR prediction model used in the present study for the southeast coastal area of China is as follows:

(1). The model is used when the predicted typhoon track is judged as potentially having an impact on rainfall in the target area (i.e., when the distance between the typhoon center and the stations is less than 500 km).
(2). According to the historical typhoon tracks, the top 27 typhoons that are most similar to the track of the target typhoon are selected.
(3). The TWIC equation is used to correct the TAR according to the typhoon intensity for selected typhoons observed at 75 stations.
(4). The TAR of selected typhoons is averaged after intensity correction.

The spatial distribution of the RMSE (mm) and correlation coefficients of 55 typhoons at 75 stations in the eastern and southern coastal areas of China from 1961 to 2017 estimate using this TAR prediction model during the training period are presented in Figure 5. Here, the RMSE of the 55 typhoons at the majority of stations is seen to be below 70 mm. The particularly large error and low correlation at the southern and eastern coastlines of China may be due to the impact of the co-existing rainy front in southeastern China and the relatively strong TC passing through the coastal areas, since a strong rainfall intensity with considerable regional variations can reduce the accuracy of TAR predictions.

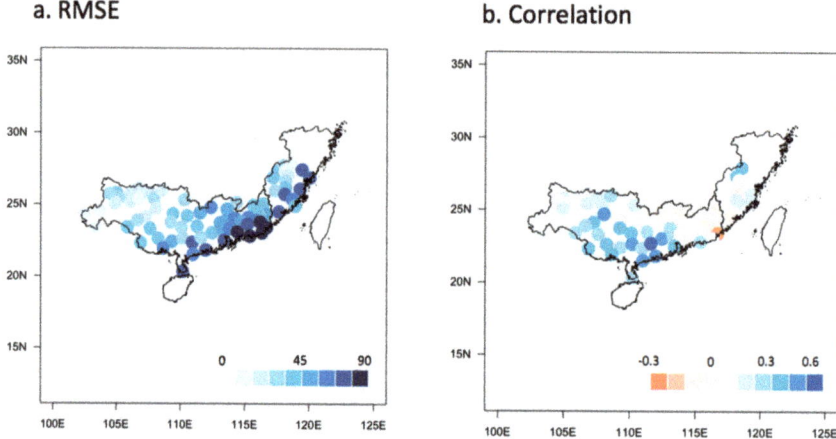

Figure 5. Graphs showing the average RMSE (**a**) and correlation coefficient (**b**) calculated using the prediction model for 55 typhoons at 75 stations during the model training period.

3.3. Model Performance

Typhoon Sarika (#1621), which affected the coastal area of southern China in 2016, typhoon Nesat (#1709), which affected the coastal area of southeast China in 2017, and typhoon Utor (#0104), which passed between Hainan Island and Taiwan, were then used to evaluate the actual performance of the TAR prediction model. The three typhoons had different tracks as they approached and made landfall in China, with Sarika (#1621) crossing Hainan Island and moving northwestward to land along the southern coastline of Guangxi Province in China, while Nesat (#1709) landed in Fujian Province through the Taiwan Strait after passing through northern Taiwan and then moving southwest. The FCM approach was used to selected typhoons with the most similar tracks, then their TAR intensities were corrected according to the aforementioned equation and were then averaged. The most similar tracks obtained from the FCM analysis are presented in Figure 6. By averaging the TC wind intensity-corrected historical TAR records of these typhoons, the TAR values of Sarika (#1621), Nesat (#1709), and Utor (#0104) at the 75 stations in the southeastern coastal and southern coastal areas of China were predicted and compared with the observed values. The results indicate RMSE values of 35.7, 55.5, and 47.2 mm for typhoons Nesat (#1709), Utor (#0104), and Sarika (#1621), respectively. Thus, the error in the results of TAR prediction for two of the three typhoons using the proposed statistical model is lower than the average error (51.2 mm) obtained using 55 typhoons during the model training period.

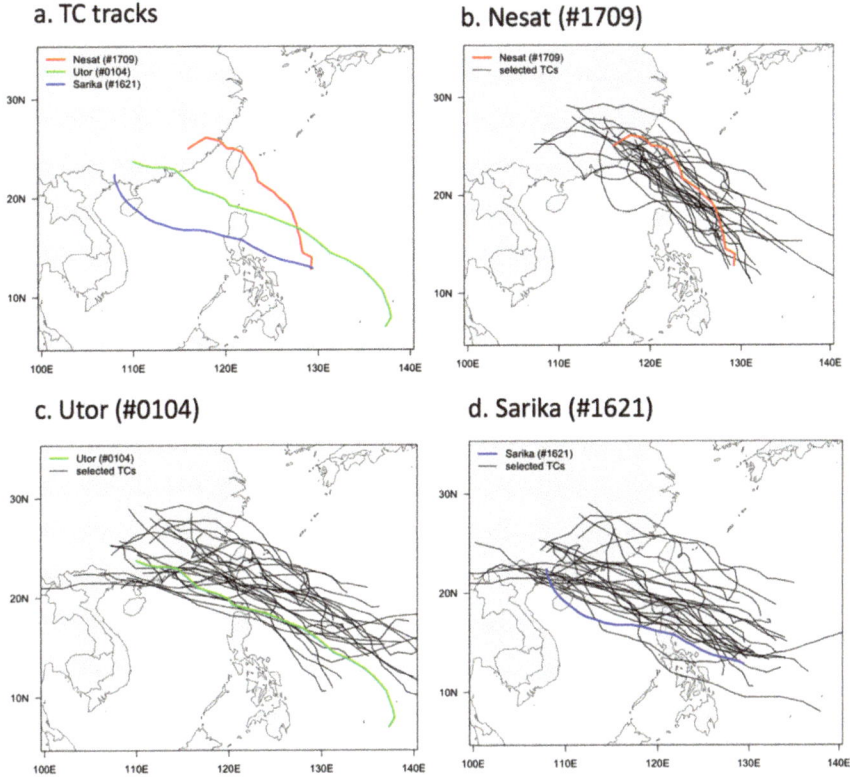

Figure 6. Verification of the prediction model established in the present study: (**a**) the TC tracks used; (**b**–**d**) the selected typhoon trajectories most similar to those of (**b**) typhoon Nesat (#1709); (**c**) typhoon Utor (#0104); and (**d**) typhoon Sarika (#1621). The typhoon numbers and similarity levels are indicated in the key.

The observed TAR values for typhoons Nesat (#1709), Utor (#0104), and Sarika (#1621) are presented, along with the differences between observed and predicted values, in Figure 7. Here, the predicted TAR spatial pattern for typhoon Nesat (#1709) is seen to be very similar to the observed outcome, except that the TAR for part of the area farther away from the coast is overestimated. For typhoon Utor (#0104), the predicted results show significant differences in the Southeastern and Pearl river basins, being slightly overestimated in the former and underestimated in the latter compare to the observation. For typhoon Sarika (#1621), the distribution of predicted TAR values in the southern and southeastern coastal areas of China is very similar to the actual observations, although it is overestimated in Fujian Province and underestimated in Guangdong Province. These results are further illustrated by the violin plots (boxplot-density trace synergism) in Figure 8. In conclusion, the results of the TAR prediction model presented in this study are effectively similar to the actual observations and indicate the overall good performance of the model for predicting the spatial distribution of TAR values.

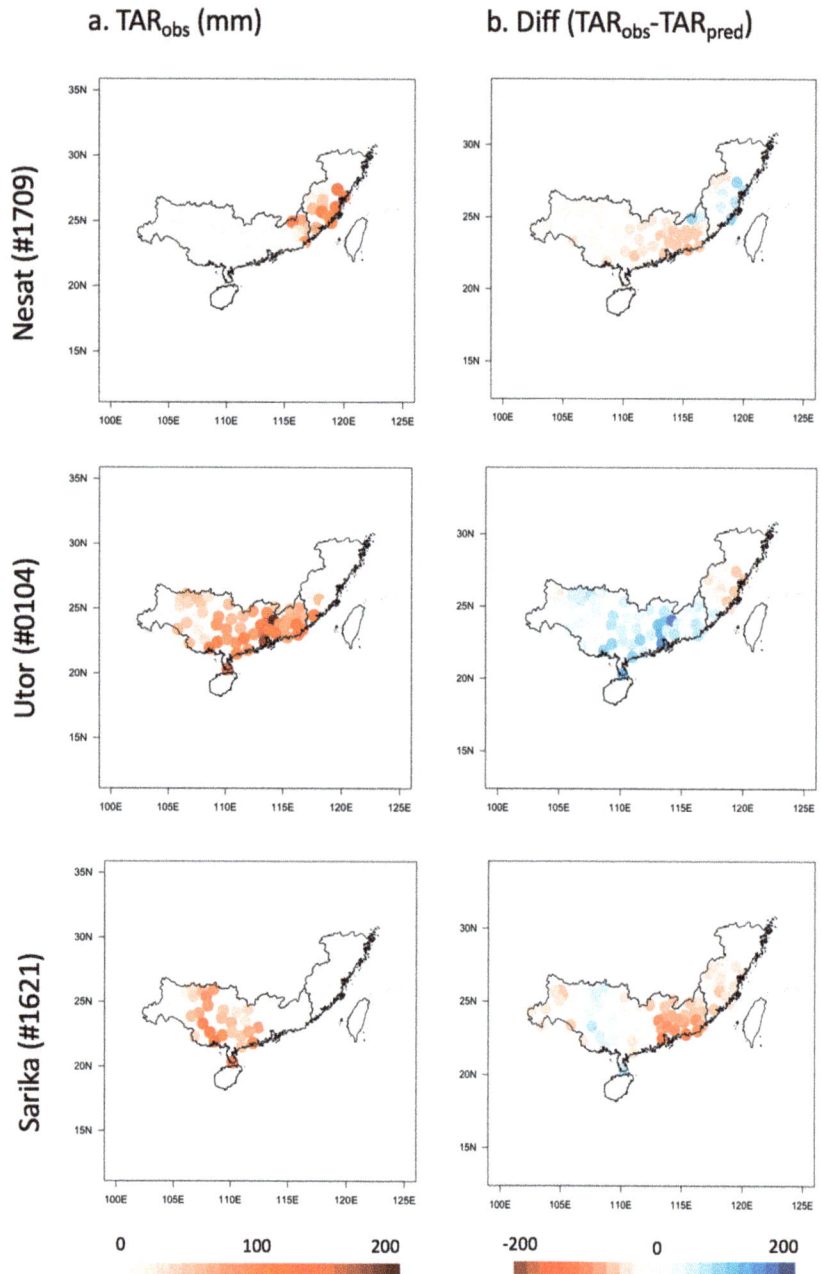

Figure 7. TAR estimation at 75 stations along the southern and southeastern coasts of China for typhoons Nesat (#1709), Utor (#0104), and Sarika (#1621): (**a**) the observed values; and (**b**) the difference between observed and predicted values.

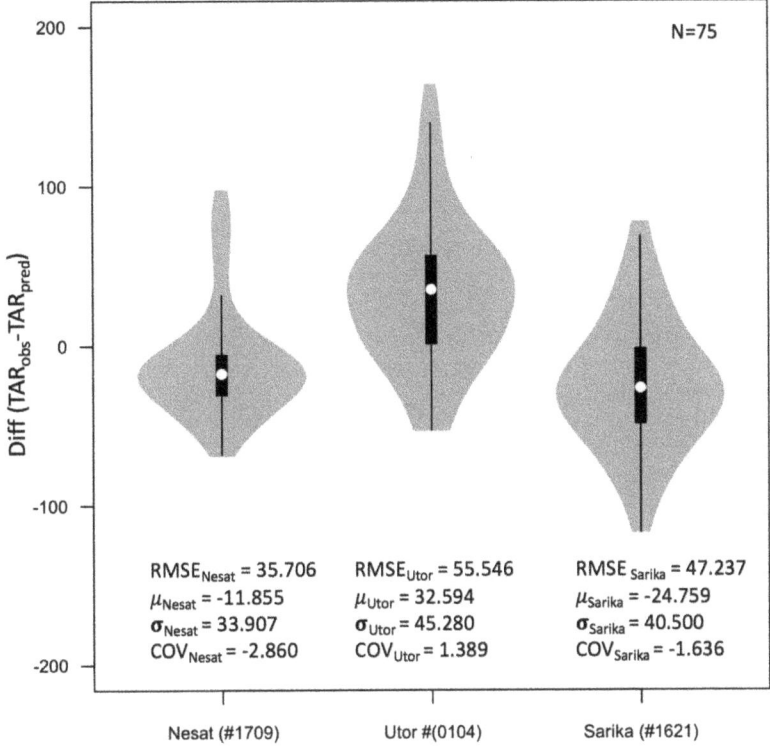

Figure 8. Violin plots (boxplot-density trace synergism) of the TAR difference between the observed and predicted values for 75 stations along the southern and southeastern coasts of China for typhoons Nesat (#1709), Utor (#0104), and Sarika (#1621). The white circles indicate the median value of the TAR difference for 75 stations.

4. Summary and Conclusions

A statistical approach for predicting typhoon rainfall was developed herein based on the historical storm track, intensity, and rainfall data for 55 typhoons affecting the southeastern coastal areas of China from 1961 to 2017. Specifically, the statistical model was based on the principle of track similarity. Since tropical cyclones (TCs) with similar tracks tend to produce relatively similar rainfall patterns, therefore, historical TC rainfall data with similar tracks were used to predict the accumulated rainfall caused by the target TC. In addition, TC intensity correction and ensemble averaging for multiple similar TC tracks were used to reduce prediction errors. The fuzzy C-means clustering (FCM) algorithm was used to select the typhoons with the most similar tracks to that of the target typhoon. The typhoon-induced accumulated rainfall (TAR) values of the selected typhoons observed at each of the 75 stations were corrected according to typhoon intensity, and then averaged to provide an estimate of the target typhoon's TAR value at each station.

The results indicated an average error of 51.2 mm across the 75 stations in the coastal area of southern China. In addition, three typhoons that were excluded from the model training process (i.e., Nesat (#1709), Utor (#0104), and Sarika (#1621)) were subsequently used to generate a forecast according to their best-track data and, thus, verify the predictive performance of the model. The resulting RMSE for the predicted TAR of Utor (#0104) is slightly high (55.5 mm), while those of Nesat (#1709) and Sarika (#1621) were 35.7 and 47.2 mm, respectively. The latter two errors were lower than the average error (51.2 mm) obtained during the model training period, thus proving the feasibility of the model for use in actual predictions. Subsequently, the spatial distribution results of the TAR values for the three typhoons predicted by this model at 75 stations were analyzed and found to be similar to the actual observations. This further demonstrated the overall good performance of the model in predicting the spatial distribution of the TAR values.

Nevertheless, the TAR prediction model presented in this study is limited to predicting only the accumulated rainfall caused by typhoons; it cannot predict the change in rainfall over time at all locations. Although numerical weather prediction (NWP) models are more advanced in this respect, the results predicted by the proposed statistical model have greater significance in certain contexts—especially for regulating reservoir discharge and flood control. The roles of the proposed model are to provide a more accurate forecast of the TAR at the target site, to coordinate the prediction of traditional numerical models, and to ensure that the region has responded well to typhoon-related rainfall measures.

Predicting rainfall caused by typhoons is challenging because, in addition to the track and intensity of the typhoon, many factors such as the regional terrain, the interaction of the typhoon with the land, and the speed of the storm translation can have certain effects upon the TAR. Notably, the TAR prediction model established in the present study did not consider these factors. Additionally, the number of typhoon samples used to build the TAR prediction model in the southern and southeastern coastal areas of China was not large. If additional factors are considered in future research, such as a correction for storm translation speed and size, and if the effective sample size is increased by using more typhoon data, the predicted results might become more accurate. In addition, confirmation is required via a comparison with NWP-based ensemble prediction models. All these approaches can help improve the performance of the TAR prediction model over China.

Author Contributions: Conceptualization, Resources, Formal analysis, Writing—Original draft, J.-S.K., and A.C.; Conceptualization, Methodology; I.-J.M., J.L., and J.-S.K., Writing—Review & editing, I.-J.M., Y.-I.M. and J.-S.K. All authors have read and agreed to the published version of the manuscript.

Funding: We appreciate the support of the State Key Laboratory of Water Resources and Hydropower Engineering Science, Wuhan University.

Conflicts of Interest: The authors declare no conflict of interest.

References

1. McCarthy, J.J. *IPCC, Climate Change 2001: Impacts, Adaptation and Vulnerability: Contribution of Working Group II to the Third Assessment Report of the Intergovernmental Panel on Climate Change*, 1st ed.; Cambridge University Press: Cambridge, UK, 2001; ISBN 0521015006.
2. Neumann, B.; Vafeidis, A.T.; Zimmermann, J.; Nicholls, R.J. Future Coastal Population Growth and Exposure to Sea-Level Rise and Coastal Flooding—A Global Assessment. *PLoS ONE* **2015**, *10*, e0118571. [CrossRef] [PubMed]
3. Kang, H.Y.; Kim, J.S.; Kim, S.Y.; Moon, Y.I. Changes in High- and Low-Flow Regimes: A Diagnostic Analysis of Tropical Cyclones in the Western North Pacific. *Water Resour. Manag.* **2017**, *31*, 3939–3951. [CrossRef]
4. Kim, J.S.; Jain, S.; Kang, H.Y.; Moon, Y.I.; Lee, J.H. Inflow into Korea's Soyang Dam: Hydrologic variability and links to typhoon impacts. *J. Hydro Environ. Res.* **2019**, *22*, 50–56. [CrossRef]
5. Kim, J.S.; Kim, S.T.; Wang, L.; Wang, X.; Moon, Y.I. Tropical cyclone activity in the northwestern Pacific associated with decaying Central Pacific El Niños. *Stoch. Env. Res. Risk A* **2016**, *30*, 1335–1345. [CrossRef]
6. Knuston, T.R.; McBride, J.L.; Chan, J.; Emanuel, K.; Holland, G.; Landsea, C.; Held, I.; Kossin, J.P.; Srivastava, A.K.; Sugi, M. Tropical cyclones and climate change. *Nat. Geosci.* **2010**, *3*, 157–163.
7. Wang, Y.; Shen, X.S.; Chen, D.H. Verification of tropical cyclone rainfall predictions from CMA and JMA global models. *J. Trop. Meteorol.* **2012**, *18*, 537.
8. Chen, P.; Yu, H.; Chan, J.C.L. A western North Pacific tropical cyclone intensity prediction scheme. *Acta Meteorol. Sin.* **2011**, *25*, 611–624. [CrossRef]
9. Kim, J.S.; Jain, S. Precipitation trends over the Korean peninsula: Typhoon-induced changes and a typology for characterizing climate-related risk. *Environ. Res. Lett.* **2011**, *6*, 034033. [CrossRef]

10. Rogers, R.; Marks, F.; Marchok, T. Tropical cyclone rainfall. In *Encyclopedia of Hydrological Sciences*; John Wiley & Sons, Ltd.: Chichester, UK, 2009; ISBN 9780471491033. [CrossRef]
11. Lian, Y.; Liu, Y.; Dong, X. Strategies for controlling false online information during natural disasters: The case of Typhoon Mangkhut in China. *Technol. Soc.* **2020**, *62*, 101265. [CrossRef]
12. Kim, T.K. *Impacts of Data Assimilation on the K-PPM for Hydro-Meteorological Application: Based on Analysis of Typhoon Cases Landed on the Korean Peninsula during 2012 (Unpublished Doctoral Dissertation)*; Pukyong National University: Busan, Korea, 2013.
13. Bender, M.A.; Ginis, I.; Tuleya, R.E.; Thomas, B. The operational GFDL coupled Hurricane–Ocean prediction system and a summary of its performance. *Mon. Weather Rev.* **2007**, *135*, 3965–3989. [CrossRef]
14. Lin, I.I.; Wu, C.C.; Pun, I.F.; Ko, D.S. Upper-ocean thermal structure and the western North Pacific category 5 typhoons. Part I: Ocean features and the category 5 typhoons' intensification. *Mon. Weather Rev.* **2008**, *136*, 3288–3306. [CrossRef]
15. Elsberry, R.L.; Carr, L.E., III. Consensus of dynamical tropical cyclone track forecasts errors versus spread. *Mon. Weather Rev.* **2000**, *128*, 4131–4138. [CrossRef]
16. Goerss, J.S. Tropical cyclone track forecasts using an ensemble of dynamical models. *Mon. Weather Rev.* **2000**, *128*, 1187–1193. [CrossRef]
17. Gao, S.; Zhao, P.; Pan, B.; Li, Y.; Zhou, M.; Xu, J.L.; Zhong, S.; Shi, Z. A nowcasting model for the prediction of typhoon tracks based on a long short term memory neural network. *Acta Oceanol. Sin.* **2018**, *37*, 8–12. [CrossRef]
18. Kim, W.J.; Hasegwa, O. Time Series Prediction of Tropical Storm Trajectory Using Self-Organizing Incremental Neural Networks and Error Evaluation. *J. Adv. Comput. Intell. Intell. Inform.* **2018**, *22*, 465–474. [CrossRef]
19. Chen, G.M.; Yu, H.; Cao, Q.; Zeng, Z.H. The performance of global models in TC track forecasting over the western North Pacific from 2010 to 2012. *Tropical Cyclone Res. Rev.* **2013**, *2*, 149–158.
20. Lei, X.T.; Wong, W.K.; Fong, C. A Challenge of the Experiment on Typhoon Intensity Change in Coastal Area. *Tropical Cyclone Res. Rev.* **2017**, *6*, 94–97.
21. Mackey, B.P.; Krishnamurti, T.N. Ensemble forecast of a typhoon flood event. *Weather Forecast.* **2001**, *16*, 399–415. [CrossRef]
22. Kidder, S.Q.; Knaff, J.A.; Kusselson, S.J.; Turk, M.; Ferraro, R.R.; Kuligowski, R.J. The tropical rainfall potential (TRaP) technique. Part I: Description and examples. *Weather Forecast.* **2005**, *20*, 456–464. [CrossRef]
23. Marks, F.D.; Kappler, G.; DeMaria, M. Development of a tropical cyclone rainfall climatology and persistence (RCLIPER) model. In *Reprints of the 25th Conference on Hurricanes and Tropical Meteorology*; American Meteorological Society: San Diego, CA, USA, 2002.
24. Lonfat, M.; Rogers, R.; Marchok, T.; Marks, F.D., Jr. A parametric model for predicting hurricane rainfall. *Mon. Weather Rev.* **2007**, *135*, 3086–3097. [CrossRef]
25. Liu, C.C. The influence of terrain on the tropical rainfall potential technique in Taiwan. *Weather Forecast.* **2009**, *24*, 785–799. [CrossRef]
26. Li, Q.; Lan, H.; Chan, J.C.L.; Cao, C.; Li, C.; Wang, X. An operational statistical scheme for tropical cyclone-induced rainfall forecast. *J. Trop. Meteorol.* **2016**, *21*, 101–110.
27. Ebert, E.E.; Turk, M.; Kusselson, S.J.; Yang, J.; Seybold, M.; Keehn, P.R.; Kuligowski, R.J. Ensemble tropical rainfall potential (e TRaP) forecasts. *Weather Forecast.* **2011**, *26*, 213–224. [CrossRef]
28. Kim, H.J.; Moon, I.J.; Kim, M. Statistical prediction of typhoon-induced accumulated rainfall over the Korean Peninsula based on storm and rainfall data. *Meteorol. Appl.* **2019**, 1–18. [CrossRef]
29. Ren, F.; Qiu, W.; Ding, C.; Jiang, X.; Wu, L.; Xu, Y.; Duan, Y. An objective index of tropical cyclone track similarity and its preliminary application in the prediction of the precipitation associated with landfalling tropical cyclones. *Weather Forecast.* **2018**, *33*, 1725–1742. [CrossRef]
30. Lee, C.-S.; Huang, L.-R.; Shen, H.-S.; Wang, S.-T. A climatology model for forecasting typhoon rainfall in Taiwan. *Nat. Hazards* **2006**, *37*, 87–105. [CrossRef]
31. Wu, Y.; Wu, S.; Zhai, P. The impact of tropical cyclones on Hainan Island's extreme and total precipitation. *Int. J. Climatol.* **2007**, *27*, 1059–1064. [CrossRef]
32. Son, C.-Y.; Kim, J.-S.; Moon, Y.-I.; Lee, J.-H. Characteristics of tropical cyclone-induced precipitation over the Korean River basins according to three evolution patterns of the Central-Pacific El Niño. *Stoch. Env. Res. Risk A.* **2017**, *28*, 1147–1156. [CrossRef]

33. Dhakal, N.; Jain, S. Nonstationary influence of the North Atlantic tropical cyclones on the spatio-temporal variability of the eastern United States precipitation extremes. *Int. J. Climatol.* **2019**, *40*, 3486–3499. [CrossRef]
34. Bezdek, J.C.; Ehrlich, R.; Full, W. FCM: The fuzzy c means clustering algorithm. *Comput. Geosci.* **1984**, *10*, 191–203. [CrossRef]
35. Kim, H.S.; Kim, J.H.; Ho, C.H.; Chu, P.S. Pattern classification of typhoon tracks using the fuzzy c-means clustering method. *J. Climate.* **2011**, *24*, 488–508. [CrossRef]
36. Cerveny, R.S.; Newman, L.E. Climatological relationships between tropical cyclones and rainfall. *Mon. Weather Rev.* **2000**, *128*, 3329–3336. [CrossRef]
37. Qi, L.; Yu, H.; Chen, P. Selective ensemble-mean technique for tropical cyclone track forecast by using ensemble prediction systems. *Q. J. Roy. Meteor. Soc.* **2014**, *140*, 805–813. [CrossRef]

Publisher's Note: MDPI stays neutral with regard to jurisdictional claims in published maps and institutional affiliations.

© 2020 by the authors. Licensee MDPI, Basel, Switzerland. This article is an open access article distributed under the terms and conditions of the Creative Commons Attribution (CC BY) license (http://creativecommons.org/licenses/by/4.0/).

Article

Seasonal Precipitation Variability and Non-Stationarity Based on the Evolution Pattern of the Indian Ocean Dipole over the East Asia Region

Jong-Suk Kim [1], Sun-Kwon Yoon [2,*], Sang-Myeong Oh [3] and Hua Chen [1]

[1] State Key Laboratory of Water Resources and Hydropower Engineering Science, Wuhan University, Wuhan 430072, China; jongsuk@whu.edu.cn (J.-S.K.); chua@whu.edu.cn (H.C.)
[2] Department of Safety and Disaster Prevention Research, Seoul Institute of Technology, Seoul 03909, Korea
[3] Operational Systems Development Department, National Institute of Meteorological Sciences, Jeju 63568, Korea; sicilia@korea.kr
* Correspondence: skyoon@sit.re.kr

Citation: Kim, J.-S.; Yoon, S.-K.; Oh, S.-M.; Chen, H. Seasonal Precipitation Variability and Non-Stationarity Based on the Evolution Pattern of the Indian Ocean Dipole over the East Asia Region. *Remote Sens.* **2021**, *13*, 1806. https://doi.org/10.3390/rs13091806

Academic Editor: Christopher Kidd

Received: 23 March 2021
Accepted: 29 April 2021
Published: 6 May 2021

Publisher's Note: MDPI stays neutral with regard to jurisdictional claims in published maps and institutional affiliations.

Copyright: © 2021 by the authors. Licensee MDPI, Basel, Switzerland. This article is an open access article distributed under the terms and conditions of the Creative Commons Attribution (CC BY) license (https://creativecommons.org/licenses/by/4.0/).

Abstract: Non-linear behavioral links with atmospheric teleconnections were identified between the Indian Ocean Dipole (IOD) mode and seasonal precipitation over East Asia (EA) using statistical models. The analysis showed that the lower the lag time, the higher the correlation; more than a two-fold correlation for non-linear regression with a kernel density estimator than for the linear regression method. When the IOD peaked, a pattern of significant reductions in seasonal precipitation during the negative IOD period occurred throughout the Korean Peninsula (KP). The occurrence of the positive IOD was in line with the El Niño phenomenon and generated greater seasonal precipitation than only the positive IOD, which takes place from March to May. This change occurred more in the cold tongue El Niño than the warm pool El Niño, inducing much higher spring precipitation throughout the KP. When negative IODs and La Niña coincided, there was slightly greater precipitation from March to May compared to the sole occurrence of negative IODs. In positive (negative) IOD years, there was anti-cyclonic (cyclonic) circulation in the South China Sea (SCS), helping to transport moisture to EA. The composite precipitation anomalies in the positive (negative) IOD years show above (below) normal precipitation in southern China. In contrast, other parts of the EA experienced drier (humid) signals than normal years. In positive IOD years, the anti-cyclonic circulation strength of the Bay of Bengal and the SCS continued until autumn and spring of the following year. This shows possible remote connections between climate events related to the tropical Indian Ocean and variations in precipitation over EA.

Keywords: Indian Ocean Dipole mode; El Niño–Southern Oscillation; singular spectrum analysis; mutual information; non-stationarity of seasonal precipitation

1. Introduction

The frequency and intensity of extreme climate events have gradually increased; this has been attributed to rising global temperatures [1–3]. Seasonal variations in regional water resource availability are also closely linked to the characteristic changes in global climate [2,4–7]. These trends have significant implications for the efficient prediction and management of available water resources. It is increasingly important to understand the relationship between extreme climatic events and the seasonal variability of water resources using hydro-meteorological variables.

Long-term hydro-meteorological changes are highly correlated with large-scale atmospheric teleconnections that predict the behavior of non-linear climate systems using ocean-related climate indices, such as the El Niño–Southern Oscillation (ENSO) and the Indian Ocean Dipole (IOD) mode [8–12]. Many studies on ENSO and the IOD report the shared understanding that these systems are major sources of large-scale atmospheric environmental changes. These systems are also closely correlated with seasonal variations, such as precipitation and streamflow within local patterns of hydro-meteorological change [10–15].

The IOD mode defined by Saji et al. [10] is characterized by extreme rainfall and wet conditions in the East African region during positive IOD (p-IOD) years. During negative IOD (n-IOD) years, wet conditions typically occur in the western part of the Indian Ocean and Indonesia; this is directly impacting East Africa, triggering dry conditions. Several studies have argued that the IOD phenomenon currently precedes ENSO events evidenced by the rise in sea surface temperature (SST), leading the latter by three to six months [16,17]. However, correlation analyses between SST anomalies of ENSO and IOD in the tropical ocean region indicates that the IOD is a phenomenon that occurs independently of ENSO, and is an internal mode within the Indian Ocean region. The results from past research also demonstrate that the Indian Ocean SST is experiencing changes to its trends over time, affecting the cycle, intensity, and genesis of the IOD mode [10,18–21].

A previous study on the potential mechanism of IOD patterns over the East Asian (EA) region presented the Bonin high formation mechanism during August for the deep ridge near Japan [22]. This theory was based on the hypothesis that an equivalent-barotropic ridge near Japan was formed because of the upper troposphere (Silk Road pattern). Furthermore, Guan and Yamagata [23] suggested that the IOD event was closely related to teleconnections around Japan, Korea, and the northeastern part of China during the sweltering and dry summer of 1994. They determined that the monsoon–desert mechanism [24], producing dry conditions over the EA region, connects a Rossby wave source with IOD-induced heating around the Bay of Bengal. The upper troposphere propagates northeastward from southern China, and the Rossby wave pattern influences precipitation changes over EA. Zhang et al. [25] examined the effects of the IOD on summer precipitation over eastern China. They found that IOD forcing in the preceding autumn has a pronounced, albeit delayed, influence on the precipitation in the following summer, particularly over the Yangtze-Huaihe River Valley. Weng et al. [26] discovered a possible link between the Indian Ocean SSTA pattern and summer precipitation in China by anomalous mid- and low-level tropospheric circulations. Cai et al. [27] identified the IOD impact on Australian winter rainfall using the Rossby wave train.

IOD and ENSO patterns are considered the main causes of large-scale atmospheric change, leading to significant changes in the hydro-meteorological patterns of several EA countries [12,22,23,28–30]. Recent studies have suggested that global surface temperature rise may have to slow down due to significant heat transfer from the Pacific to Indian Oceans via the Indonesian Throughflow [31–33]. Studies on Indo-Pacific thermocouples are required to improve the current understanding of climatic variability and quantitatively diagnose such variability at a regional scale. Existing quantitative studies on the characteristics of the IOD and ENSO phenomena, relating to Korean watersheds and their regional assessments, are relatively inadequate. This study analyzed the influence of long-term precipitation variability in the EA region by examining p-IOD and n-IOD events [10]. The classification of IOD events in accordance with the patterns of p-IOD and n-IOD events was also analyzed, and the analysis of the evolution patterns of the IOD was based on the approach of Saji and Yamagata [20]. This study addresses three specific objectives: (1) to analyze the significant changes in large-scale pattern and long-term precipitation variability in the EA, and in the KP region sub-watershed using ocean-related, abnormal climate phenomena, in accordance with the IOD mode index; (2) to investigate linkages between atmospheric teleconnections and possible mechanisms between different phases of the IOD and seasonal precipitation over the KP using statistical methods; and (3) to carry out a diagnostic study on the non-stationarity and possibility of seasonal precipitation prediction, using climate indices during significant IOD and ENSO seasons over the KP.

2. Materials and Methods

2.1. Data

SST data obtained from the National Center for Environmental Prediction and the National Center for Atmospheric Research (NCEP-NCAR) reanalysis v2 were used to classify the IOD events in the Tropical Indian Ocean (TIO). IOD Mode Index (DMI) data were obtained from the Japan Agency for Marine-Earth Science and Technology (JAMSTEC) and the Hadley Centre Sea Ice and Sea Surface Temperature dataset. The DMI is defined as the SST anomaly difference between the western (50°E–70°E, 10°S–10°N) and southeastern (90°E–110°E, 10°S–equator) regions of the TIO region [10]. The DMI calculated by the Asia-Pacific Economic Cooperation Climate Center (APCC), Busan, South Korea, was used; this utilizes monthly SST from the National Oceanic and Atmospheric Administration (NOAA) Extended Reconstructed Sea Surface Temperature (ERSST) v4 in the TIO region. The Global Precipitation Climatology Center (GPCC) monthly precipitation, which has a regular grid with a spatial resolution of 0.5° × 0.5° latitude by longitude from Deutscher Wetterdienst in Germany was used to diagnose precipitation variability and its long-term changes in the EA region. In addition, spatially averaged daily precipitation data provided by the Water Resources Management Information System (WAMIS) were used to undertake a detailed examination of the regional impact in five major Korean river basins in the KP (Figure 1). The average precipitation was calculated using the Thiessen polygon network from 125 precipitation gauge stations for 117 sub-watersheds within these five major river basins foMIr 1966–2016. For the composite analysis, monthly vector wind anomalies at 850 hPa were used; these were obtained from the NCEP-NCAR. Student's t-test was used for statistical significance testing for composite analysis.

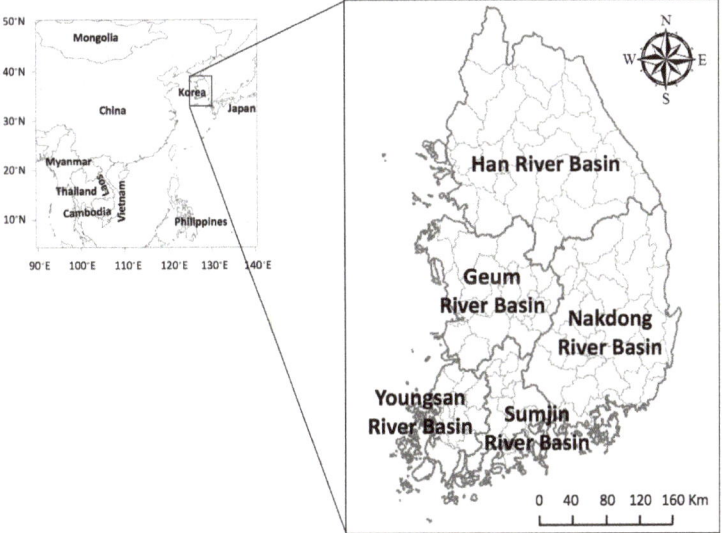

Figure 1. Map of the East Asia region, including the location of the five major river basins of the Korean Peninsula.

2.2. Classification of IOD Events

This study used the methodology proposed by Saji and Yamagata [20] to classify positive and negative IOD events in the TIO region. The process included data pre-processing and exclusion of data that did not meet the following criteria:

(1) Pre-processing of data: SST anomalies in the Western Indian Ocean (10°S–10°N, 60°–80°E) and eastern Indian Ocean (10°S–0°, 90°–110°E), and zonal wind anomalies over the equator (U_{eq}, area-averaged wind anomaly over 5°S–5°N, 70°–90°E), were

first detrended. A three-month running mean was then applied once over the three time-series datasets to reduce the impact of intra-seasonal fluctuations;
(2) Identifying criteria: The DMI and U_{eq} needed to exceed 0.5 σ in amplitude for at least three months. In addition, the SSTA in the west and east Indian Ocean have opposite signs, and the magnitude should exceed 0.5 σ for at least three months.

Figure 2 shows the DMI region in the Indian Ocean and the normalized anomaly time series for the DMI between the Hadley Centre Sea Ice and the Sea Surface Temperature dataset (HadISST) by the NOAA Climate Prediction Center (CPC), and the method proposed by Saji et al. [10]. Two time-series datasets were compared using a similar method; the 13 strongest p-IOD, and the 15 strongest n-IOD years were classified based on the HadISST data from 1956 to 2018.

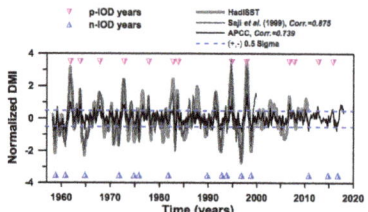

Figure 2. The dipole mode index (DMI) region in the Indian Ocean and normalized anomaly time series for DMI. (**a**) DMI region in the Indian Ocean, (**b**) normalized anomaly time series of the DMI. In the panel of the (**b**), the light gray, dark gray, and solid black lines indicate calculated DMI indices by HadISST, Saji et al. [10] and APEC Climate Center (APCC), respectively. "corr" indicates the correlation coefficient.

2.3. Singular Spectrum Analysis

Singular spectrum analysis (SSA) is a non-parametric spectral estimation method that reduces dispersion by changing the coordinates in time series through techniques derived from principal component analysis (PCA); this enables the extraction of information from noisy time series. By removing the non-harmonic components from the original time series data, the long-term frequency and trend may easily be understood [34]. SSA embeds the data of a time series $X_i (1 < i < N)$ in a vector space of dimension M, and applies the empirical orthogonal function (EOF) method. This enables the projection of original data in the orthogonal functions EOF 1 and EOF 2. By composing the axes using these EOFs, the trend, cycle, and tendency of total variance in the data, are more clearly apparent.

To separate the non-harmonic components, the size of the eigenvalues was defined by an orthogonal process. The orthogonal function was calculated between α_i^1 and α_i^2; these are the orthogonal coefficients of the principal component (PC1) and PC2 time series that correspond with the harmonic components in the original coordinates, X and Y. Finally, it was converted to the reconstruction component (RC) as R_i^1 and R_i^2. The estimation forecasting model may be configured to reflect specific characteristics such as frequency and trend in the original data. By using Equation (1) to reconstruct the data, the original data may be replaced with a new time series with a constant frequency and less noise:

$$\begin{aligned}
(R_A X)_i &= \tfrac{1}{i} \sum_{j=1}^{i} \sum_{k \in A} \alpha_{i-j}^k E_j^k, \ 1 \leq i \leq M-1 \\
(R_A X)_i &= \tfrac{1}{M} \sum_{j=1}^{M} \sum_{k \in A} \alpha_{i-j}^k E_j^k, \ M \leq i \leq N-M+1 \\
(R_A X)_i &= \tfrac{1}{N-i+1} \sum_{j=1-N+M}^{i} \sum_{k \in A} \alpha_{i-j}^k E_j^k, \ N-M+2 \leq i \leq N
\end{aligned} \quad (1)$$

where α_i^k is an orthogonal coefficient; E_j^k is the empirical orthogonal function ($1 \leq k \leq M$); M indicates a dimension; and τ is the sampling rate. Additional detailed information is available from Moon and Lall [35].

2.4. Mutual Information

Mutual information (*MI*) is one of the most popular measures that determines the extent to which one random variable (*Y*) may communicate information on another random variable (*X*). This may also be considered an exercise in reducing uncertainty on one random variable given some knowledge of another variable. This is a useful tool to calculate non-linear correlations between different datasets. Non-linear correlations indicate that the ratio of change between variables is not constant. Here, *MI* was used to extract information regarding the non-linear correlation between climate indices and seasonal precipitation in the KP. In general, the nonlinear correlation is high or low as the sum of *MI* values quantitatively represents the correlation. The *MI* method has a conditional occurrence probability by section. If the *MI* is large, the non-linear correlation between the two datasets is also large. Where appropriate lag times are selected, the *MI* technique may also be used to estimate the probability density function using a kernel function in a non-parametric manner.

If there are two types of time series datasets, such as ($s_1, s_2, s_3, \cdots, s_n, q_1, q_2, q_3, \cdots, q_n$); where *n* is the observed period, then the *MI* between observations s_i and q_j is defined by Equation (2) [35]:

$$MI_{s,q}(s_i, q_j) = log_2 \left(\frac{P_{s,q}(s_i, q_j)}{P_s(s_i) P_q(q_j)} \right) \quad (2)$$

where $P_{s,q}(s_i, q_j)$ indicates the joint probability density function between *s* and *q*, calculated by a time series of (s_i, q_j), and $P_s(s_i)$ and $P_q(q_j)$ are the marginal probability densities calculated from s_i and q_j, respectively. The average mutual information ($I_{s,q}$) of the two discrete random variables *s* and *q* can be defined using Equation (3):

$$I_{s,q} = \sum_{i,j} P_{s,q}(s_i, q_j) log_2 \left(\frac{P_{s,q}(s_i, q_j)}{P_s(s_i) P_q(q_j)} \right) \quad (3)$$

where $P_{s,q}(s_i, q_j)$ is the joint probability distribution function of *X* and *Y*, and $P_s(s_i)$ and $P_q(q_j)$ are the marginal probability distribution functions of *s* and *q*, respectively. This equation is useful to determine whether the components in multivariate sampling are independent or dependent. In particular, Martinerie et al. [36] and Gao and Zheng [37] used *MI* techniques to construct a state space for appropriate lag time selection in an orthogonal time series.

The *MI* analysis between the two datasets was performed using Equation (4), proposed by Joe [38], following the standard normal distribution of the axis (*X, Y*) and its linear correlation analysis:

$$I(X; Y) = -0.5 \, log\left[1 - \rho(X, Y)^2\right] \quad (4)$$

where $I(X; Y)$ indicates the calculated average *MI* value through *MI* analysis, and $\rho(X, Y)$ is the linear correlation between *X* and *Y*.

MI based on the non-linear correlation coefficient may be used to obtain $\lambda [0 \leq \lambda \leq 1]$. To calculate λ by estimating the average *MI* value following the standard normal distribution in the two variables *X* and *Y*, Equation (5) proposed by Joe [38] and Granger and Lin [39] was used:

$$\hat{\lambda}(X, Y) = \sqrt{1 - exp\left[-2\hat{I}(X, Y)\right]} \quad (5)$$

where $\hat{I}(X, Y)$ is the average *MI* value from the two variables *X* and *Y* and $\hat{\lambda}(X, Y)$ is a non-linear correlation coefficient estimated from the average *MI* value between the two variables (*X, Y*). In this study, a linear regression (LR) method using Equation (5) was used with the estimated average *MI* values and non-linear regression using Equation (5) with two-dimensional Kernel density estimators (KDE) [34]. The 95% confidence limits were estimated using 1000 bootstrap resampling replications, enabling more accurate calculation of the confidence limits, given the limited data. The advantages of the *MI* and SSA used in this study are highly suitable to capture non-parametric relationships from data without imposing structures or restrictions on the model. SSA helps to identify similar spectral components in two or more time series, which may be interpreted as connections between these series. However, wavelet coherence considers two time series. SSA is primarily driven by data, while wavelet analysis may be influenced by the selection of the parent wavelet function [40].

In general, abnormal SSTs in the TIO region may have triggering effects on the troposphere temperature rise due to enhanced air–sea interaction [41–43]. Atmospheric teleconnection links with the jet stream can affect variations in local precipitation worldwide, even in the EA region [44,45]. The effects of precipitation variability over the KP and EA regions due to changes in IOD patterns were diagnosed. Although they are geographically remote areas, they may affect and hydrologically correlate by atmospheric-dynamic processes and associated mechanisms [10,20]. The p-IOD and n-IOD events were analyzed from April (when developing had commenced), through September (when it peaked), and up to November, where it had begun to disappear. This study analyzed the impact of IOD evolution patterns on the KP within a three-month window, accommodating for a one-month delay. To understand the role of IOD events in atmospheric variability over the KP, linear and non-linear correlations were analyzed, along with the lag time correlations between the IOD and local precipitation variations.

3. Analysis and Results

3.1. Nonlinear Atmospheric Teleconnections over the KP

The non-linear lag time correlations were calculated using MI, and their lag-time correlations were simulated from lag-0 to lag-11 (Figures 3 and 4). Figure 3 shows the joint probability kernel density functions among the normalized three-month moving average precipitation and p-/n-IOD indices over the KP. The result of the joint probability kernel density function is based on the *MI* results for lag-1 month non-linear correlations. For the precipitation of the KP, the probable mode values corresponding to the vertices of the joint probability density function were 0.632, 0.603, and 0.601 at lag times of 1, 3, and 6, respectively; the probable mode values tended to decrease with respect to lag time. There was a positive correlation between seasonal precipitation and IOD pattern changes in each lag time over the KP. This result was based on the analysis of the location of central points of the joint probability kernel density functions with precipitation and the IOD index, using *MI* techniques.

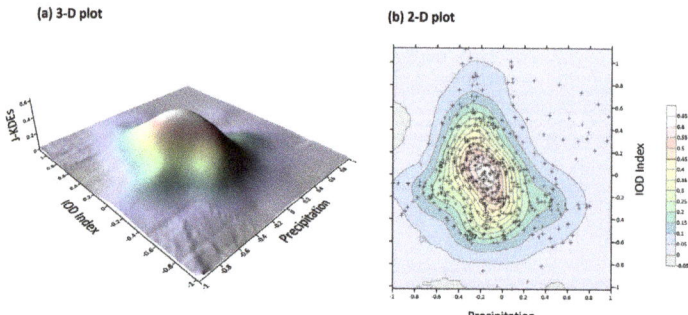

Figure 3. Lag-1 month nonlinear correlation of joint probability kernel density estimations (J-KDEs) between the normalized three-month moving average precipitation and IOD index in the Korean Peninsula.

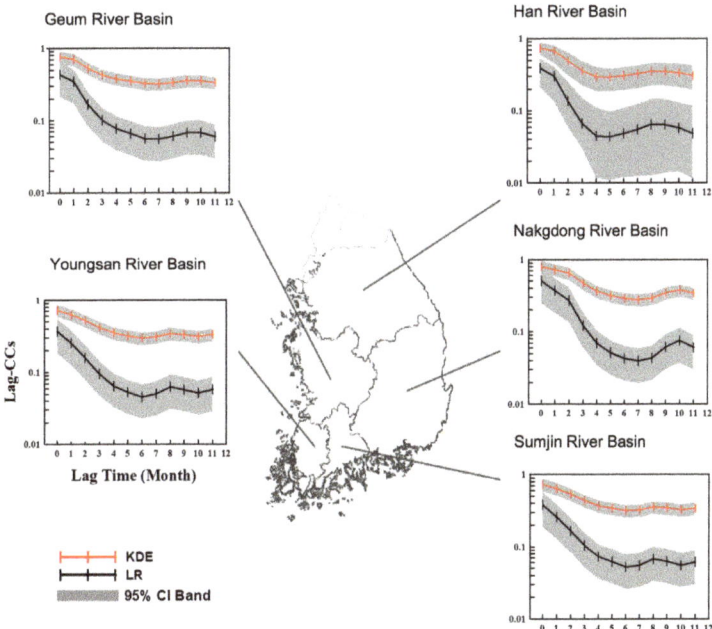

Figure 4. Non-linear and linear lag-time correlation coefficients (CCs) with 95% confidence limits between IOD index and monthly precipitation for the five major rivers of the Korean Peninsula. Confidence limits are given by 5% and 95% quantiles of 1000 bootstrap resampling replications.

Figure 4 presents the linear and non-linear correlation coefficients (CCs) with their 95% confidence limits between climate indices and precipitation for the five major Korean rivers using the KDE and LR approaches. The lag-0 correlation had the highest correlation for LR and KDE, which was correlated with the IOD index and KP precipitation (LR: 0.315, KDE: 0.684). These time lags indicate a non-linear correlation between climate indices and monthly precipitation; as such, there is a possibility for a diagnostic study on the seasonal or sub-seasonal prediction of local precipitation over the KP using ocean-related large-scale climate indices.

3.2. Evolution Pattern of the Indian Ocean Dipole and Its Local Impacts over the KP

Figure 5 presents analysis results for the change in precipitation in the KP according to the evolution pattern of the p-IOD years from April to November. Total precipitation in the KP decreased significantly from the long-term average, with −11.90% in April–June, −8.63% in May–July, −14.32% in June–August, −9.92% in July–September, −15.23% in August–October, and −7.14% in September–November. The total amount of precipitation change was analyzed using Student's t-test; it was found that there was a significant decrease in the southern part of the KP at a 95% confidence level. For the p-IOD phases, the pattern of decreased precipitation was more likely to occur at a significant level in the southern part than the mid-northern part of the KP. The changes in this pattern persisted significantly between April and November in the p-IOD years. During the August–October period (autumn in Korea), a distinct pattern of decreased precipitation was observed mainly in the central and southern KP.

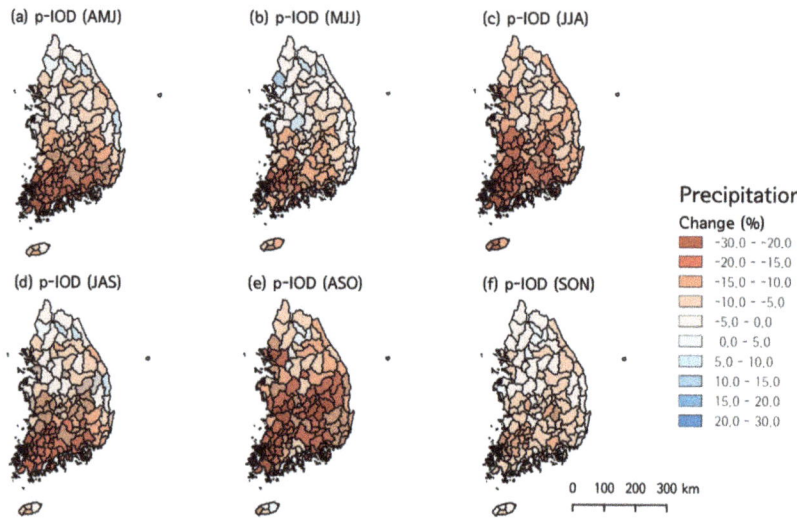

Figure 5. Evolution pattern of the seasonal precipitation in the KP in the Indian Ocean Dipole mode during p-IOD years. Hatched polygons indicate statistically significant changes of season precipitation at a 5% significance level. Each figure shows seasonal precipitation anomalies over three months, from April to November.

Figure 6 presents the results from the analysis of changes in precipitation in the KP according to the evolution pattern of the n-IOD years from April to November. Total precipitation in the KP tended to decrease or increase more than usual, with −0.71% in April–June, +7.91% in May–July, −1.39% in June–August, −4.05% in July–September, 7.89% in August–October, and −1.07% in September–November. A pattern of significant decreased precipitation in the central part of the KP was observed. However, a pattern of increased precipitation occurred in the southern part of the KP during May–July. For the n-IOD phases, contrary to the p-IOD phases, the pattern of significant decreased precipitation was more likely to appear in the northern KP, as opposed to the central or southern KP. These changes appear to be conspicuous between April and November, when an n-IOD was observed. During p-IOD events (Figure 7a,c), the annual/June–September precipitation in the KP was −7.14%/−14.74% lower than the long-term average annual/June–September precipitation (1971–2000). During n-IOD events (Figure 7b,d), the annual/June–September precipitation in the KP slightly decreased to −1.07%/−3.31%. The composite analysis revealed that the June–September precipitation during p-IOD events was substantially lower than that during long-term normal years. In contrast, n-IOD events had annual/June–September precipitation that was slightly below normal conditions.

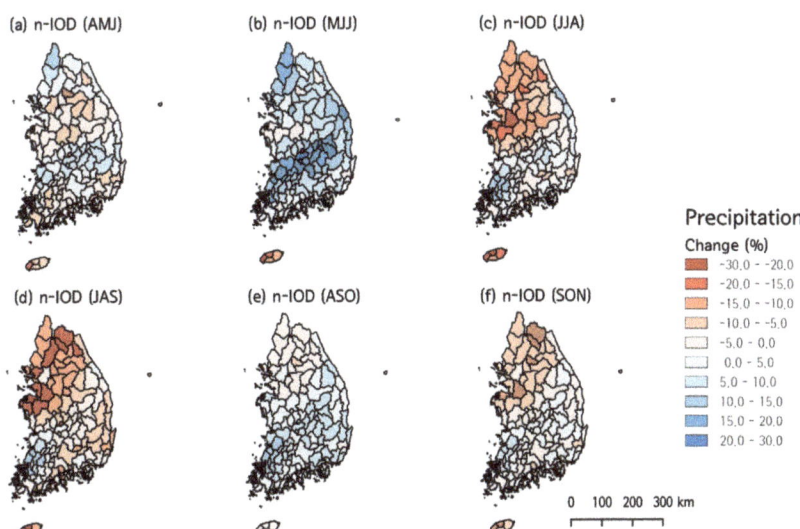

Figure 6. Evolution pattern of the seasonal precipitation in the KP in the Indian Ocean Dipole mode during n-IOD years. Hatched polygons indicate statistically significant changes of season precipitation at a 5% significance level. Each figure shows seasonal precipitation anomalies over three months, from April to November.

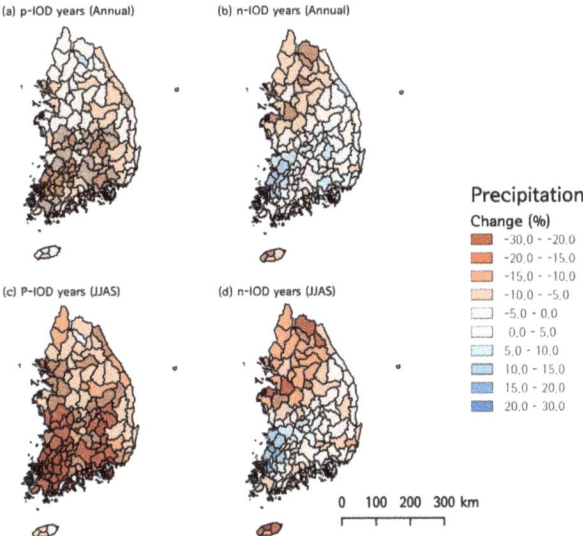

Figure 7. Composite anomalies of annual and JJAS (June–September) precipitation during p-/n-IOD years. Hatched polygons indicate statistically significant changes in precipitation at a 5% significance level.

The mechanisms of major climate phenomena associated with tropical oceans, such as ENSO and IOD, are not yet fully understood because it is still a challenge to simulate them completely using physical climate models of the global environment. In addition, analyzing and predicting climate phenomena through physical models involves considerable difficulties. Based on the physical model results, applying them to hydrologic circulation systems in a specific EA region, such as KP and China, may follow the problem of scientific reliability and understanding. Therefore, this study analyzed non-linear behavior links with atmospheric teleconnections between hydro-meteorological variables and the climate index using statistical models over the KP with the ocean-related major climate indices, including ENSO and IOD. Statistical approaches have the disadvantage of making it difficult to expect significant levels of results because of the limited number of observations. However, it is one of the most important methods that can be used to complement the prediction results of physical models.

3.3. Nonstationarity of Seasonal Precipitation Anomalies for Different Phases of the IOD

Figures 8 and 9 show changes in the 30-year mean precipitation in five major rivers of the KP. In each figure, Case I shows change over time in the 30-year mean precipitation without excluding the effects of the IOD. Case II is the result of excluding the precipitation in the p-IOD years, and Case III excludes the precipitation in the n-IOD years.

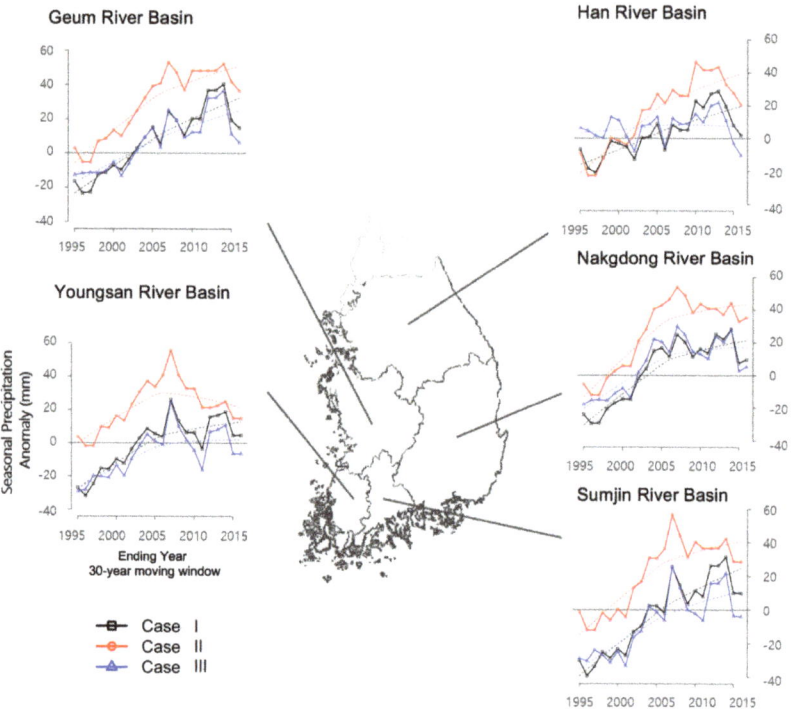

Figure 8. Changes in seasonal precipitation anomalies (August–October) when the IOD peaked. In each panel, Case I shows the changes over time in the 30-year mean precipitation without excluding the effects of IOD separately. Case II is the result of excluding the precipitation in the positive IOD years, and Case III shows the change in the 30-year mean precipitation, excluding the precipitation in the n-IOD years.

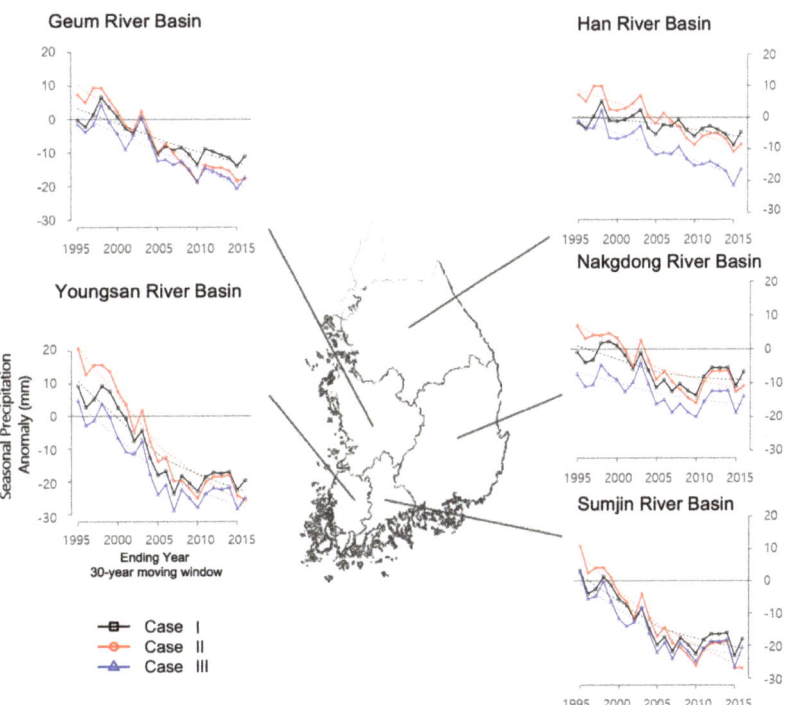

Figure 9. Changes in seasonal precipitation anomalies (March–May) of the following years when the IOD peaked. In each panel, Case I shows the changes over time in the 30-year mean precipitation without excluding the effects of IOD separately. Case II is the result of excluding the precipitation in the p-IOD years, and Case III shows the change in the 30-year mean precipitation, excluding the precipitation in the n-IOD years.

Figure 8 illustrates the seasonal precipitation from August to October when the IOD peaked; this precipitation from August to October in all five major rivers had a statistically significant increase ($p < 0.001$). This change in precipitation occurred in the Han River Basin and was relatively abundant in the southern part of the KP. In the Youngsan River Basin, Cases I and III showed a statistically significant increase in the 30-year mean precipitation analysis. Although there was an increase in seasonal precipitation, this change was not statistically significant ($p > 0.05$). In the Han River Basin, Cases I and II showed statistically significant increases; in contrast, in Case III, there was an increase in seasonal precipitation, although it was not statistically significant ($p > 0.05$). As shown in the GPCC composite analysis, the p-IOD years in the Youngsan River Basin led to reduced precipitation from the long-term normal throughout the KP. This was particularly the case in the central part of the KP, where a precipitation reduction occurred in the Han River Basin. Notably, the central river basins of the KP (Han and Geum River Basins) have experienced a sharp decline in seasonal precipitation since 2013, and the southern basins have tended to shift from increased seasonal precipitation to declining or plateauing patterns since 2007.

Figure 9 shows changes in the 30-year mean precipitation in March–May, when the IOD crosses the peak and enters a period of decline. The IOD showed a statistically significant decline through the basins of the five rivers, in contrast to precipitation during the peak IOD season. The decline in seasonal precipitation from March–May was noticeable in the Youngsan and Sumjin River Basins in the southern coastal region of Korea.

At times, the IOD co-occurs with ENSO; as such, in this study, the effects of IOD and ENSO on seasonal precipitation changes were analyzed when the IOD peaked, and when the IOD and ENSO entered a period of decline. The effect of the combination of IOD and ENSO is shown in the composite precipitation analysis (Table 1). For the p-IOD phases in the August to October period, the pattern of significant decreases in precipitation was more likely to appear over the entire KP. In contrast, during the p-IOD years, the entire KP experienced precipitation that was 2.6–9.4% greater than the average precipitation in March–May. The occurrence of p-IOD coincided with El Niño events, resulting in more seasonal precipitation in March–May than the occurrence of only p-IODs. This occurred more frequently with cold tongue (CT) El Niño than warm pool (WP) El Niño, resulting in significantly greater spring precipitation across the KP. For the n-IOD years, there was lower precipitation than usual in the Han River Basin (7.4%), followed by the Sumjin River Basin (5.2% reduction in March–May precipitation), and the Youngsan River Basin (3.4% reduction in March–May precipitation). When n-IOD co-occurred with La Niña, there was slightly greater precipitation in March–May than for n-IOD in isolation. These findings indicate that IOD events strongly influence precipitation and its sub-watersheds in the KP. This shows a linkage of possible teleconnections and characteristic changes between tropical Indian-Ocean-related major climactic events and local precipitation variability over the KP.

Table 1. Changes in seasonal precipitation from the long-term normal (1966–2016) (unit: %).

River Basin	August–October		March–May			
	p-IOD Years	n-IOD Years	p-IOD Years	n-IOD Years	p-IOD/El Niño	n-IOD/La Niña
Han River	−11.0	0.8	6.5	−7.4	7.0 (20.5)	−1.5
Nakdong River	−26.8	−2.0	2.6	2.3	13.6 (20.4)	8.9
Geum River	−24.6	4.3	4.0	3.9	11.8 (22.0)	−1.5
Sumjin River	−26.0	5.2	3.1	−5.2	18.8 (25.1)	7.4
Youngsan River	−25.9	6.0	9.4	−3.4	13.3 (20.6)	8.7

p-IOD years (1967, 1972, 1977, 1982, 1983, 1994, 1997, 2006, 2007, 2012, and 2015), n-IOD years (1971, 1974, 1975, 1981, 1989, 1992, 1993, 1996, 1998, 2010, 2014, and 2016). The numerical values in parentheses show the results of Case I, coinciding with CT El Niño.

3.4. Large-Scale Air–Sea Environment and Precipitation Variations over East Asia

Based on the p-IOD and n-IOD years defined above, a composite analysis of autumn (August–October) SSTA in the TIO was conducted (Figure 10). During the p-IOD years, cold SST anomaly patterns appeared in the eastern Indian Ocean, including Indonesia and the maritime continent; warm SSTA patterns emerged in the equatorial region of the Indian Ocean. The warm and cold SST distributions were not extensive, although strong signals were observed in the East Indian Ocean. Conversely, although not strongly dependent on IOD phases, there was a warm and cold SSTA distribution over large areas in the western Indian Ocean. Furthermore, in most areas where warm and cold signals appeared, the confidence level was greater than 95%.

Many recent studies have shown that these different SST anomaly patterns in the TIO region may affect air changes in circulation. Moreover, they detected several teleconnection-based significant changes in different regions of experiencing seasonal precipitation in Northeast Asia. The large-scale physical mechanism of the developing and decaying IOD phases in the atmosphere has not yet been clearly understood. However, there are several reliable studies; some are diagnostic studies, while others are regional impact assessments of the p-IOD and n-IOD [10,17–20]. After a p-IOD (n-IOD), basin-scale warming (cooling) was referred to as the Indian Ocean Basin (IOB) mode [32]. These influences on the IOB may affect the climate of EA in the following season.

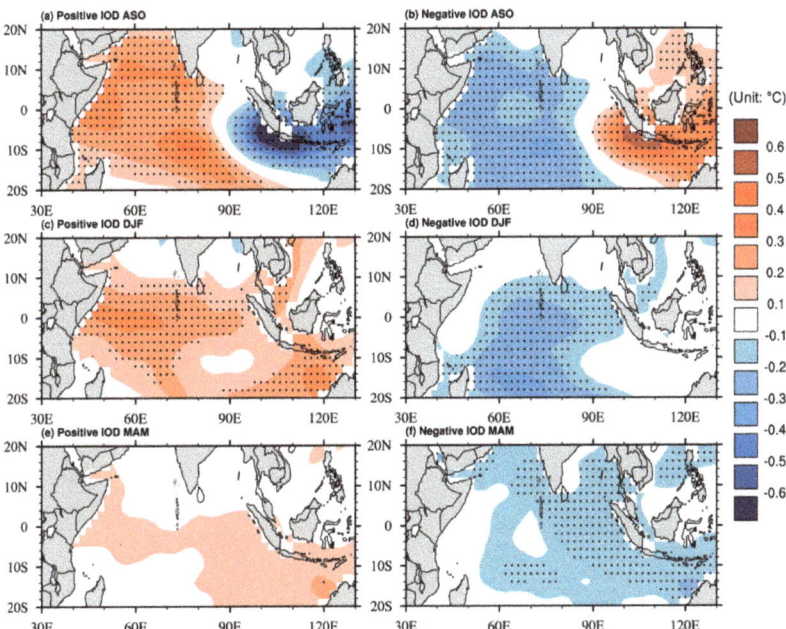

Figure 10. Composite anomalies of mean SST in the TIO region during positive and negative IOD years, from August to May in the following year. NOAA Extended Reconstructed Sea Surface Temperature version 5 (ERSSTv5) monthly data were used for the SSTA composite analysis; climatology data were used for the normal years from 1956 to 2018. Dotted points indicate values over 95% confidence based on Student's *t*-test.

Figure 11 shows the composite anomalies (1981–2010 climatology) of the GPCC precipitation and 850 hPa wind over northeast Asia during strong IOD events, in the same way as SSTA. During the warm boreal season, southwesterly winds from the Indian Ocean and the South China Sea (SCS) were dominant and advected a large amount of moisture from the Indian Ocean to EA [34]. In p-IOD (n-IOD) years, there was an anticyclonic (cyclonic) circulation in the SCS. This large circulation may help to transport (prevent) moisture to EA. The composite precipitation anomalies of p-IOD (n-IOD) years showed that they were above (below) normal over the southern parts of China. In contrast, other parts of EA, including the KP, experienced drier (wetter) signals than normal years (Figure 11a,b). In p-IOD years, southern China and the SCS are mainly affected by the southwesterly winds, and this pattern continues until the following spring. In particular, heavy precipitation occurred in southern China during the boreal winter season; this is consistent with the findings of Qui et al. [35]. During n-IOD years, easterly winds were observed in EA (Figure 11e–h).

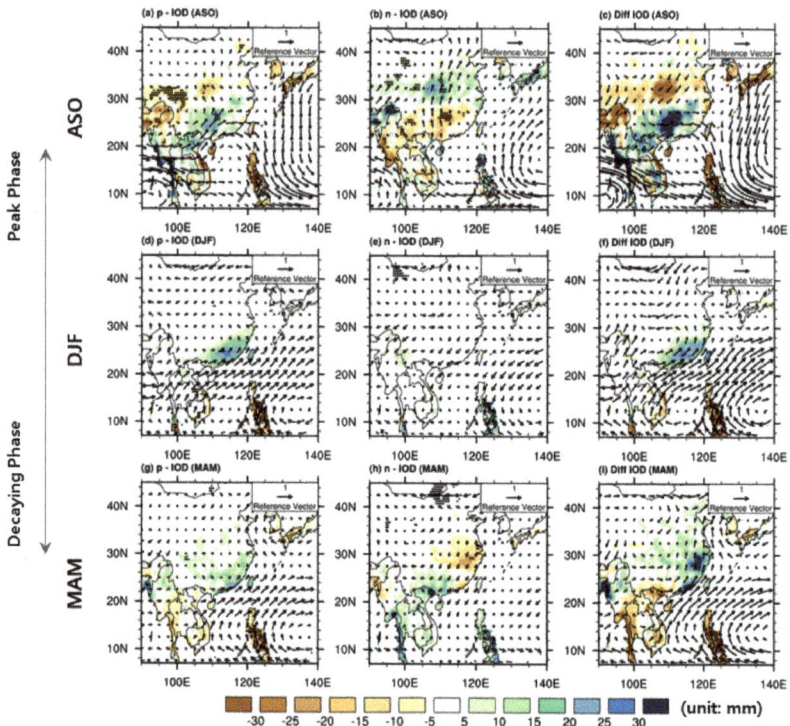

Figure 11. Composite anomalies of the GPCC precipitation and 850 hPa low-level wind anomalies from autumn (August–September–October) to spring (March–April–May) of the following year during IOD events over the northeast Asia region. The left (middle) panel indicates the p-IOD (n-IOD) events, and the right panel shows the large-scale circulation differences between p-IOD and n-IOD years. Dots denote values over 95% confidence based on Student's *t*-test.

4. Conclusions

Understanding the relationship between air–sea environments and precipitation variations in EA for areas experiencing high seasonal variability and uncertainty regarding seasonal precipitation data is critical to develop a sustainable freshwater management system. In this study, statistical models were used to analyze non-linear behavior links of atmospheric teleconnections between climate indices and seasonal precipitation. The IOD mode, a major ocean-related climatic factor in the Indian Ocean, was used to analyze long-term changes in seasonal precipitation over the EA region. The primary results are summarized as follows:

(1) The analysis of atmospheric teleconnections was conducted using PCA and SSA techniques. Non-linear lag correlations between climate indices and seasonal precipitation were calculated using the *MI* technique, and their lag-time correlations were simulated from lag-0 to lag-11. Teleconnection-based non-linear and linear CCs were conducted between climate indices and seasonal precipitation using LR and KDE based on the *MI* results. Results from non-linear CCs were higher than those from linear correlations, and IOD was found to directly influence the precipitation anomaly time series over the KP. This study demonstrates a method for teleconnection-based long-range water resource management to reduce climate uncertainty when an abnormal SSTA occurs in the TIO region;

(2) When the IOD reached its peak (August to October), a significant decrease in seasonal precipitation during the n-IOD period was observed throughout the KP. For the spring

period (March to May), seasonal precipitation during p-IOD years coincided with the El Niño phenomenon, which was higher than those of only p-IOD years. These changes occurred more frequently in the CT El Niño than in the WP El Niño years. For the co-occurrence of n-IODs and La Niña, there was greater precipitation than when only n-IODs occurred in isolation;

(3) The characteristics of non-stationary 30-year averaged seasonal precipitation were detected throughout the KP. The precipitation in autumn (August to October) was observed to increase significantly ($p < 0.001$) when excluding the p-IOD year across the KP. In contrast, seasonal precipitation in the central river basins of KP had plummeted since 2013 and decreased in the southern basins of the KP since 2007. Spring precipitation showed statistically significant declines across the five major rivers in the KP when IODs peaked and entered a period of decline. The decline in seasonal precipitation from March to May was noticeable in the southern coastal regions of Korea;

(4) During p-IOD years, there were more precipitation signals than usual in the southern part of China, including the SCS and the southern part of Japan, with cyclonic circulation patterns. A high-pressure anti-cyclonic pattern was observed over eastern China and the KP. There was a drier signal in n-IOD years than normal in the SCS and southern China, along with a high-pressure anti-cyclonic pattern. Conversely, inland and eastern regions of China and Japan showed wetter signals than usual, with a cyclonic circulation pattern. However, the KP was located between the two cyclonic circulations, and the district precipitation signal was not visible. The signal was less dry than the p-IOD years.

The results of this diagnostic study may be utilized in decision-making processes to minimize climate-related disasters, such as floods and droughts, through seasonal prediction. Additionally, these results may inform the development of optimal strategies to ensure best management practices for water use under a changing climate.

Author Contributions: Conceptualization, resources, formal analysis, writing—original draft, J.-S.K., S.-K.Y., and S.-M.O.; data curation, methodology; J.-S.K. and S.-K.Y., writing-review and editing; J.-S.K., S.-K.Y., and H.C. All authors have read and agreed to the published version of the manuscript.

Funding: The APC was funded by Seoul Institute of Technology (2020-AB-001), Seoul, South Korea, and partially supported by the Korea Meteorological Administration Research and Development Program "Development of Marine Meteorology Monitoring and Next-generation Ocean Forecasting System" under Grant (KMA2018-00420).

Data Availability Statement: The authors acknowledge the monthly SST data from the National Oceanic and Atmospheric Administration (NOAA) and the monthly vector wind anomalies at 850 hPa provided by the National Centers for Environmental Prediction and the National Center for Atmospheric Research (NCEP-NCAR) reanalysis, version 2 at https://www.ncdc.noaa.gov/data-access/model-data/model-datasets/reanalysis-1-reanalysis-2 (accessed on 5 May 2021). The monthly Indian Ocean Dipole Mode Index (DMI) data were obtained from the Japan Agency for Marine-Earth Science and Technology (JAMSTEC) at http://www.jamstec.go.jp/frsgc/research/d1/iod/iod_home.html.en (accessed on 5 May 2021), Japan, and the Hadley Centre Sea Ice and Sea Surface Temperature dataset (HadISST) by the NOAA Climate Prediction Center (CPC) at https://www.cpc.ncep.noaa.gov/ (accessed on 5 May 2021). The monthly precipitation dataset by the Global Precipitation Climatology Center (GPCC) from Deutscher Wetterdienst in Germany at https://climatedataguide.ucar.edu/climate-data/gpcc-global-precipitation-climatology-centre (accessed on 5 May 2021). The authors also acknowledged that the local precipitation data for 117 sub-watersheds of the five major river basins over the KP by the Water Resources Management Information System (WAMIS) at http://www.wamis.go.kr/ (accessed on 5 May 2021), and the multi-model ensemble climate data provided by the APEC Climate Center (APCC) at http://www.apcc21.org/ (accessed on 5 May 2021).

Acknowledgments: This work was supported by the Seoul Institute of Technology, Seoul, South Korea, and the third author, Sang-Myeong Oh, was acknowledged the Korea Meteorological Administration Research and Development Program by the Korea Meteorological Administration (KMA), South Korea. We also appreciate the support of the State Key Laboratory of Water Resources and

Hydropower Engineering Science, Wuhan University, China. The authors thank the three anonymous reviewers for their comments and valuable suggestions.

Conflicts of Interest: The authors declare no conflict of interest.

References

1. Wang, B.; Wu, R.; Fu, X. Pacific-East Asia Teleconnection: How Does ENSO Affect East Asian Climate? *J. Clim.* **2000**, *13*, 1517–1536. [CrossRef]
2. Pizarro, G.; Lall, U. El Niño-induced flooding in the U.S. West: What can we expect? *Eos Trans. Am. Geophys. Union.* **2002**, *83*, 349–352. [CrossRef]
3. IPCC (Intergovernmental Panel on Climate Change). *Managing the Risks of Extreme Events and Disasters to Advance Climate Change Adaptation (SREX)—Special Report of the Intergovernmental Panel on Climate Change*; Cambridge University Press: Cambridge, UK, 2007; pp. 1–594.
4. Horel, J.D.; Wallace, J.M. Planetary-scale atmospheric phenomena associated with the Southern Oscillation. *Mon. Weather Rev.* **1981**, *109*, 813–829. [CrossRef]
5. Kim, J.S.; Jain, S.; Yoon, S.K. Warm season streamflow variability in the Korean Han River Basin: Links with atmospheric teleconnections. *Int. J. Clim.* **2012**, *32*, 635–640. [CrossRef]
6. Yoon, S.K.; Kim, J.S.; Lee, J.H.; Moon, Y.I. Hydrometeorological variability in the Korean Han River Basin and its sub-watersheds during different El Niño phases. *Stoch. Environ. Res. Risk Assess.* **2013**, *27*, 1465–1477. [CrossRef]
7. Lee, T.S.; Ouarda, T.; Yoon, S.K. KNN-based Local Linear Regression for the Analysis and Simulation of Low Flow Extremes under Climatic Influence. *Clim. Dyn.* **2017**, *49*, 3493–3511. [CrossRef]
8. Piechota, T.C.; Dracup, J.A. Drought and regional hydrologic variation in the United States: Associations with the El Niño-Southern Oscillation. *Water Resour. Res.* **1996**, *32*, 1359–1373. [CrossRef]
9. Piechota, T.C.; Chiew, H.S.; Francis Dracup, J.A.; McMachon, T.A. Seasonal streamflow forecasting in eastern Australia and the El Niño-Southern Oscillation. *Water Resour. Res.* **1998**, *34*, 3035–3044. [CrossRef]
10. Saji, N.H.; Goswami, B.N.; Vinayachandran, P.N.; Yamagata, T. A dipole mode in the tropical Indian Ocean. *Nature* **1999**, *401*, 360–363. [CrossRef] [PubMed]
11. Yoon, S.K.; Lee, T. Investigation of hydrological variability in the Korean Peninsula with the ENSO teleconnections. *Proc. Int. Associ. Hydrol. Sci.* **2016**, *374*, 165–173. [CrossRef]
12. Kim, J.-S.; Xaiyaseng, P.; Xiong, L.; Yoon, S.-K.; Lee, T. Remote Sensing-Based Rainfall Variability for Warming and Cooling in Indo-Pacific Ocean with Intentional Statistical Simulations. *Remote Sens.* **2020**, *12*, 1458. [CrossRef]
13. Ashok, K.; Guan, Z.; Yamagata, T. Influence of the Indian Ocean dipole on the Australian winter rainfall. *Geophys. Res. Lett.* **2003**, *30*, 1821. [CrossRef]
14. McPhaden, M.J.; Zebiak, S.E.; Glantz, M.H. ENSO as an integrating concept in Earth Science. *Science* **2006**, *314*, 1740–1745. [CrossRef] [PubMed]
15. Pradhan, P.K.; Preethi, B.; Ashok, K.; Krishna, R.; Sahai, A.K. Modoki, Indian Ocean Dipole, and western North Pacific typhoons, Possible implications for extreme events. *J. Geophys. Res.* **2011**, *116*, D18108. [CrossRef]
16. Klein, S.A.; Soden, B.J.; Lau, N.C. Remote sea surface temperature variations during ENSO, Evidence for a tropical atmospheric bridge. *J. Clim.* **1999**, *12*, 917–932. [CrossRef]
17. Lau, N.C.; Nath, M.J. Atmosphere-ocean variations in the Indo-Pacific sector during ENSO episodes. *J. Clim.* **2003**, *16*, 3–20. [CrossRef]
18. Behera, S.K.; Krishnan, R.; Yamagata, T. Unusual ocean-atmosphere conditions in the tropical Indian Ocean during 1994. *Geophys. Res. Lett.* **1999**, *26*, 3001–3004. [CrossRef]
19. Webster, P.J.; Moore, A.M.; Loschnigg, J.P.; Leben, R.R. Coupled ocean–atmosphere dynamics in the Indian Ocean during 1997–98. *Nature* **1999**, *401*, 356–360. [CrossRef] [PubMed]
20. Saji, N.H.; Yamagata, T. Structure of SST and Surface Wind Variability during Indian Ocean Dipole Mode Events, COADS Observations. *J. Clim.* **2003**, *16*, 2735–2751. [CrossRef]
21. Liu, Y.; Yoon, S.-K.; Kim, J.-S.; Xiong, L.; Lee, J.-H. Changes in Intensity and Variability of Tropical Cyclones over the Western North Pacific and Their Local Impacts under Different Types of El Niños. *Atmosphere* **2021**, *12*, 59. [CrossRef]
22. Enomoto, T.; Hoskins, B.J.; Matsuda, Y. The formation of the Bonin high in August. *Q. J. R. Meteorol. Soc.* **2003**, *587*, 157–178. [CrossRef]
23. Guan, Z.; Yamagata, T. The unusual summer of 1994 in East Asia: IOD Teleconnections. *Geophys. Res. Lett.* **2003**, *30*, 1544. [CrossRef]
24. Cherchi, A.; Annamalai, H.; Masina, S.; Navarra, A.; Alessandri, A. Twenty-first century projected summer mean climate in the Mediterranean interpreted through the monsoon-desert mechanism. *Clim. Dyn.* **2016**, *47*, 2361–2371. [CrossRef]
25. Zhang, Y.; Zhou, W.; Chow, E.C.H.; Leung, M.Y. Delayed impacts of the IOD: Cross-seasonal relationships between the IOD, Tibetan Plateau snow, and summer precipitation over the Yangtze–Huaihe River region. *Clim. Dyn.* **2019**, *53*, 4077–4093. [CrossRef]
26. Weng, H.; Wu, G.; Liu, Y.; Behera, S.K.; Yamagata, T. Anomalous summer climate in China influenced by the tropical Indo-Pacific Oceans. *Clim. Dyn.* **2011**, *36*, 769–782. [CrossRef]

27. Cai, W.; van Rensch, P.; Cowan, T.; Hendon, H.H. Teleconnection pathways of ENSO and the IOD and the mechanisms for impacts on Australian rainfall. *J. Clim.* **2011**, *24*, 3910–3923. [CrossRef]
28. Kosaka, Y.; Xie, S.P. Recent global-warming hiatus tied to equatorial Pacific surface cooling. *Nature* **2013**, *501*, 403–407. [CrossRef] [PubMed]
29. Lee, S.K.; Park, W.; Baringer, M.O.; Gordon, A.L.; Huber, B.; Liu, Y. Pacific origin of the abrupt increase in Indian Ocean heat content during the warming hiatus. *Nat. Geosci.* **2015**, *8*, 445–449. [CrossRef]
30. Liu, W.; Xie, S.P.; Lu, J. Tracking ocean heat uptake during the surface warming hiatus. *Nat. Commun.* **2016**, *7*, 10926. [CrossRef]
31. Zheng, X.T.; Xie, S.P.; Liu, Q. Response of the Indian Ocean Basin Mode and Its Capacity Effect to Global Warming. *J. Clim.* **2012**, *24*, 6146–6164. [CrossRef]
32. Qu, X.; Huang, G. An Enhanced Influence of Tropical Indian Ocean on the South Asia High after the Late 1970s. *J. Clim.* **2012**, *25*, 6930–6941. [CrossRef]
33. Roxy, M.K.; Gnanaseelan, C.; Parekh, A.; Chowdary, J.S.; Singh, S.; Modi, A.; Kakatkar, R.; Mohapatra, S.; Dhara, C.; Shenoi, S.C.; et al. Indian Ocean Warming. In *Assessment of Climate Change over the Indian Region*; Krishnan, R., Sanjay, J., Gnanaseelan, C., Mujumdar, M., Kulkarni, A., Chakraborty, S., Eds.; Springer: Singapore, 2020. [CrossRef]
34. Moon, Y.I.; Lall, U. Atmospheric flow indices and interannual Great Salt Lake variability. *J. Hydrol. Eng.* **1996**, *1*, 55–62. [CrossRef]
35. Moon, Y.I.; Rajagopalan, B.; Lall, U. Estimation of mutual information using kernel density estimators. *Phys. Rev. E* **1995**, *52*, 2318–2321. [CrossRef] [PubMed]
36. Martinerie, J.M.; Albano, A.M.; Mees, A.I.; Rapp, P.E. Mutual Information, Strange Attractors, and the Optimal Estimation of Dimension. *Phys. Rev. A* **1992**, *45*, 7058–7064. [CrossRef] [PubMed]
37. Gao, J.; Zheng, Z. Direct dynamical test for deterministic chaos and optimal embedding of a chaotic time series. *Phys. Rev. E* **1994**, *49*, 3807. [CrossRef]
38. Joe, H. Relative entropy measures of multivariate dependence. *J. Am. Stat. Assoc.* **1989**, *84*, 157–164. [CrossRef]
39. Granger, C.; Lin, J. Using the mutual information coefficients to identify lags in nonlinear models. *J. Time. Ser. Anal.* **1994**, *15*, 371–384. [CrossRef]
40. Chavez, M.; Cazelles, B. Detecting dynamic spatial correlation patterns with generalized wavelet coherence and non-stationary surrogate data. *Sci. Rep.* **2019**, *9*, 7389. [CrossRef]
41. Yu, B.; Lupo, A.R. Large-Scale Atmospheric Circulation Variability and Its Climate Impacts. *Atmosphere* **2019**, *10*, 329. [CrossRef]
42. Perlwitz, J.; Knutson, T.; Kossin, J.P.; LeGrande, A.N. Large-scale circulation and climate variability. In *Climate Science Special Report: Fourth National Climate Assessment, Volume I*; Wuebbles, D.J., Fahey, D.W., Hibbard, K.A., Dokken, D.J., Stewart, B.C., Maycock, T.K., Eds.; U.S. Global Change Research Program: Washington, DC, USA, 2017; pp. 61–184. [CrossRef]
43. Yang, J.; Liu, Q.; Xie, S.-P.; Liu, Z.; Wu, L. Impact of the Indian Ocean SST basin mode on the Asian summer monsoon. *Geophys. Res. Lett.* **2007**, *34*, L02708. [CrossRef]
44. Yuan, Y.; Yang, H.; Zhou, W.; Li, C. Influences of the Indian Ocean dipole on the Asian summer monsoon in the following year. *Int. J. Clim.* **2008**, *28*, 1849–1859. [CrossRef]
45. Qui, Y.; Cai, W.; Guo, X.; Ng, B. The asymmetric influence of the positive and negative IOD events on China's rainfall. *Sci. Rep.* **2014**, *4*, 4943. [CrossRef]

Article

Remote Sensing-Based Rainfall Variability for Warming and Cooling in Indo-Pacific Ocean with Intentional Statistical Simulations

Jong-Suk Kim [1], Phetlamphanh Xaiyaseng [1], Lihua Xiong [1], Sun-Kwon Yoon [2] and Taesam Lee [3],*

[1] State Key Laboratory of Water Resources and Hydropower Engineering Science, Wuhan University, Wuhan 430072, China; jongsuk@whu.edu.cn (J.-S.K.); lar99@yahoo.com (P.X.); xionglh@whu.edu.cn (L.X.)
[2] Department of Safety and Disaster Prevention Research, Seoul Institute of Technology, Seoul 03909, Korea; skyoon@sit.re.kr
[3] Department of Civil Engineering, ERI, Gyeongsang National University, 501 Jinju-daero, Jinju, Gyeongnam 660-701, Korea
* Correspondence: tae3lee@gnu.ac.kr

Received: 11 April 2020; Accepted: 3 May 2020; Published: 4 May 2020

Abstract: This study analyzed the sensitivity of rainfall patterns in South China and the Indochina Peninsula (ICP) using statistical simulations of observational data. Quantitative changes in rainfall patterns over the ICP were examined for both wet and dry seasons to identify hotspots sensitive to ocean warming in the Indo-Pacific sector. The rainfall variability was amplified by combined and/or independent effects of the El Niño–Southern Oscillation and the Indian Ocean Dipole (IOD). During the years of El Niño and a positive phase of the IOD, rainfall is less than usual in Thailand, Cambodia, southern Laos, and Vietnam. Conversely, during the years of La Niña and a negative phase of the IOD, rainfall throughout the ICP is above normal, except in parts of central Laos, northern Vietnam, and South China. This study also simulated the change of ICP rainfall in the wet and dry seasons with intentional IOD changes and verified IOD-sensitive hotspots through quantitative analysis. The results of this study provide a clear understanding both of the sensitivity of regional precipitation to the IOD and of the potential future impact of statistical changes regarding the IOD in terms of understanding regional impacts associated with precipitation in changing climates.

Keywords: rainfall variability; Indian Ocean Dipole (IOD); El Niño–Southern Oscillation (ENSO); intentional statistical simulation

1. Introduction

Spatiotemporal variation in precipitation extremes can result from the amplification of changes in atmosphere–ocean interactions and the intensification of the hydrological cycle on both regional and global scales attributable to the effects of global climate change [1–6]. Changes in the magnitude and frequency of regional rainfall are closely related to the occurrence of floods and droughts. They have important implications not only in terms of their socioeconomic impact, but also in relation to the management of local and/or regional hydropower, irrigation, and environmental water resources [7–9]. The occurrence of extreme precipitation, which is highly likely to continue into the future, is increasingly regarded as an area of concern by the public because many countries have experienced such extreme events in recent years [7,10–13]. In particular, there has been rapid increase in both the amount of damage and the number of fatalities associated with the occurrence of extreme rainfall in developing countries because of their vulnerable infrastructure, high density of human activities, and poor land use practices and development ([14,15]).

The El Niño–Southern Oscillation (ENSO) is known for its active and predictable short-term behavior within the global climate system [16], characterized by irregular but periodic changes in the behavior of winds and sea level temperatures over the tropical eastern Pacific Ocean. Since the 2000s, new forms of El Niño have appeared more frequently in the central Pacific [17,18]. However, little is yet known about the causes of these new types of El Niño, some of which have been reported to have a noticeable effect on the supply of warm seasonal freshwater and hydrological extremes in Pacific Rim countries [5,19–22]. Research over the past two decades has identified a distinct climate anomaly in the Indian Ocean, known as the Indian Ocean Dipole (IOD) [23–26]. The IOD is an atmosphere–ocean coupling mode characterized by the opposition of anomalies of sea surface temperature (SST) in the west and east of the tropical Indian Ocean [23,24,27]. A positive (negative) IOD pattern is characterized by water warmer (cooler) than normal in the western tropical Indian Ocean (10° S–10°N, 50°–70°E) and water cooler (warmer) than normal in the southeastern tropical Indian Ocean (10° S to the equator, 90°–110°E). These events usually begin around May or June and they terminate rapidly in early winter after reaching a peak between August and October [24]. Long-term climatic change has high correlation with large-scale atmospheric teleconnections and it has been reported to be predictable in relation to the behavior of nonlinear climate systems, particularly in terms of ocean-related climatic drivers such as ENSO and the IOD mode [23,24]. ENSO and IOD patterns are known as leading causes of large atmospheric change and they are related closely to seasonal variations in precipitation in the Indian Ocean region and around the world [18,28–30].

Recent studies have suggested that the observed slowdown in the rise of global mean surface atmospheric temperature is closely related to the considerable transport of heat from the Pacific Ocean into the Indian Ocean via the Indonesian Throughflow [31–33]. Investigation of Indo-Pacific thermocouples can help both to improve understanding of the regional-scale climatic variability that is globally relevant and to diagnose quantitatively such variability in a changing climate. However, there has been little previous quantitative research on rainfall variation across the Indochina Peninsula (ICP) in relation to IOD phenomena and ENSO evolution. Therefore, based on historical observations, this study undertook quantitative analysis of the changes in SST in the Indo-Pacific sector and the associated interseasonal variation of precipitation over the ICP. The study had three primary areas of interest: (1) the spatiotemporal changes in magnitude and frequency of precipitation during the dry and wet seasons, (2) the relationship between the changes in weather extremes and large-scale climatic patterns over the ICP, and (3) identification of IOD-sensitive hotspots using the intentionally biased bootstrapping (IBB) technique based on limited historical observations.

2. Materials and Methods

2.1. Precipitation Dataset and Climate Change Indices

This study used the high-resolution (0.5° × 0.5°) daily Climate Prediction Center Global Unified Precipitation dataset for 1979–2018, which was obtained from the website of NOAA's Earth System Research Laboratory's Physical Research Division (https://www.esrl.noaa.gov/psd/). The Global Precipitation Climatology Center monthly precipitation dataset with 1.0° × 1.0° spatial resolution for the period 1948–2018, which is based on quality-controlled data from 67,200 stations worldwide [34], was also used to identify seasonal precipitation variability over the ICP region (5°–25°N, 90°–115°E) (Figure 1). To identify changes in the frequency and intensity of rainfall, six major climate change indices [35], based on the daily Climate Prediction Center data from 1979–2018, were analyzed for both the wet season (May–October) and the dry season (November–April). These indices included the seasonal total precipitation (PRCPTOT) on wet days, seasonal total of the 95th percentile of precipitation (R95pTOT) on wet days (≥1.0 mm), seasonal maximum 1-day precipitation (RX1day), simple precipitation intensity index (SDII) with a daily precipitation amount on wet days of ≥1.0 mm, maximum number of consecutive dry days (CDD) with a daily precipitation amount of <1.0 mm, and maximum number of consecutive wet days (CWD) with a daily precipitation amount of ≥1.0 mm. Wet and dry days were calculated separately for both the wet season (May–October) and the dry season (November–April).

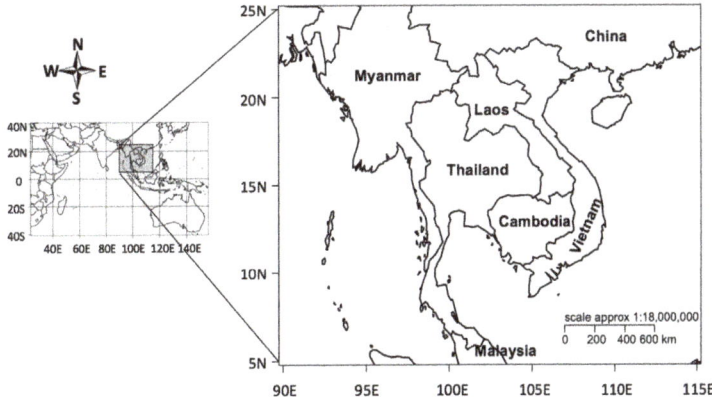

Figure 1. Map of the Indochina Peninsula (5°–25°N, 90°–115°E).

2.2. Indian Ocean Dipole (IOD) and El Niño–Southern Oscillation (ENSO)

The monthly SST anomaly (SSTA) from NOAA's Extended Reconstructed Sea Surface Temperature (ERSST) dataset v5 in the Tropical Indian Ocean (TIO) was used to calculate the IOD mode index. This is defined as the SSTA difference between the western (10° S–10°N, 50°–70°E) and southeastern (10°S to the equator, 90°–110°E) regions of the TIO [24]. From 1948–2017, a 3-month running average was applied to the IOD mode index data (August–September–October), which is the peak phase period, with ±1 SD to determine the years with positive and negative modes of the IOD (Figure 2). To characterize different types of ENSO event, monthly Niño3 (5° S–5°N, 150°E–90°W) and Niño4 (5°S–5°N, 160°E–150°W) data for the period 1948–2018 were used for El Niño development phases (December–January–February).

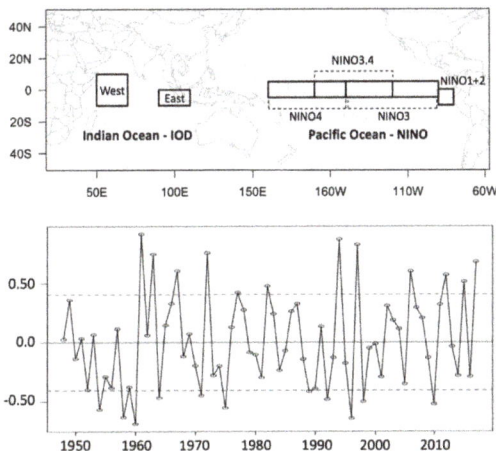

Figure 2. Dipole mode in the tropical Indian Ocean (TIO) and Niño region in the Pacific Ocean. The Indian Ocean Dipole (IOD) index is defined based on the sea surface temperature anomaly difference between the western (10° S–10°N, 50°–70°E) and southeastern (10°S to the equator, 90°–110°E) regions of the TIO shown in the upper panel. In the lower panel, the IOD time series during 1948–2017 is shown by the solid line, and the ±1 SD of the IOD is marked by dotted lines.

In this study, the pattern of El Niño was divided into two groups depending on where the peak and persistent anomalies in SST occurred in the tropical Pacific: (1) Eastern Pacific (EP), El Niño occurring in the EP; and (2) Central Pacific (CP), El Niño emerging in the CP. This study employed two new indices (Equation (1)) to identify the two types of El Niño events through a simple transformation of the Niño3 and Niño4 indices, as proposed by Ren and Jin [36]:

$$N_{CT} = N_3 - \alpha N_4 \qquad \alpha = \begin{cases} 0.4, & N_3 N_4 > 0 \\ 0, & \text{otherwise.} \end{cases} \qquad (1)$$
$$N_{WP} = N_4 - \alpha N_3,$$

Here, N_3 and N_4 indicate the Niño3 and Niño4 indices, respectively.

Assessment of the relative impacts of the IOD and ENSO on rainfall across the ICP was based mainly on composite analyses. During the period 1979–2018, the effects of ENSO and the IOD were evaluated in terms of rainfall across the ICP during both the wet season (May–October) and the dry season (November–April).

2.3. Trend Detection

A nonparametric Mann–Kendall test is commonly used to detect a monotonic pattern in a time series of climate data based on the null hypothesis that the data are independent and sorted randomly [37,38]. The null hypothesis H_0 is random in the order of the sample data (X_i, $i = 1, 2..., n$) and it has no trend, whereas the alternative hypothesis H_1 represents the monotonous tendency of X. The S statistic for Kendall's tau is calculated as follows:

$$S = \sum_{i=1}^{n-1} \sum_{j=i+1}^{n} \text{sgn}(X_j - X_i) \qquad (2)$$

and

$$\text{sgn}(\_) = \begin{cases} 1 & \text{if } \_ > 0 \\ 0 & \text{if } \_ = 0 \\ -1 & \text{if } \_ < 0 \end{cases}. \qquad (3)$$

The S statistic is calculated using the following mean and variance:

$$E(S) = 0, \qquad (4)$$

$$V(S) = \frac{n(n-1)(2n+5) - \sum_{m=1}^{n} t_m m(m-1)(2m+5)}{18} \qquad (5)$$

where t_m measures the ties of extent m. The standardized test statistic Z is estimated as follows:

$$Z = \begin{cases} \frac{S-1}{\sqrt{V(S)}} & S > 0 \\ 0 & S = 0 \\ \frac{S+1}{\sqrt{V(S)}} & S < 0 \end{cases}. \qquad (6)$$

The existence of autocorrelation in a dataset affects the probability of detecting a trend when it does not exist and vice versa, but this is often ignored. Thus, the modified nonparametric trend test developed by Hamed and Rao [39] was applied in this study. The corrected Z value is derived as follows:

$$Z = \begin{cases} \frac{S-1}{\sqrt{V^*(S)}} & S > 0 \\ 0 & S = 0 \\ \frac{S+1}{\sqrt{V^*(S)}} & S < 0 \end{cases} \qquad (7)$$

where

$$V^*(S) = V(S) * \frac{n}{n_S^*} \qquad (8)$$

$$\frac{n}{n_S^*} = 1 + \frac{2}{n(n-1)(n-2)} * \sum_{i=1}^{n-1}(n-i)(n-i-1)(n-i-2)\rho_S(i) \tag{9}$$

and where $\rho_S(i)$ is an autocorrelation function of the rank with respect to the observations. The sign of Z represents the trend direction, and the magnitude of Z is associated with the significance level, where $|Z| > 1.64$ for the 10% significance level and $|Z| > 1.96$ for the 5% significance level.

2.4. Intentionally Biased Bootstrapping Method

Bootstrapping analysis is a statistical method that can generate replicated datasets from source data, and it can evaluate the variability of their quantiles without performing separate analytical calculations [40]. However, the intentionally biased bootstrapping (IBB) technique applied in this study is a method that allows assessment of the relative effects of a response variable by deliberately increasing or decreasing the mean of the explanatory variable to a certain level while resampling it with the response variable [41].

Figure 3 shows the IBB resampling process applied in this study, and a brief description of the IBB analysis process is given below.

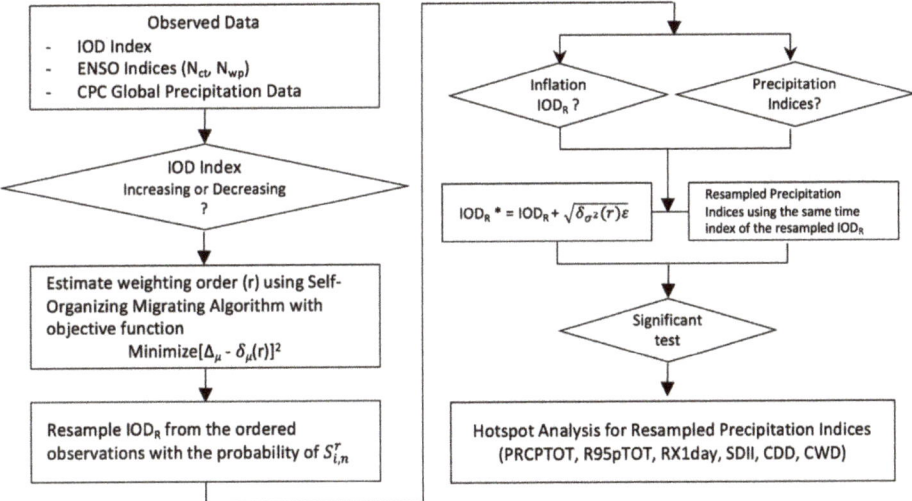

Figure 3. Procedure of the intentionally biased bootstrapping (IBB) resample analysis applied in this study.

Among n observations x_i ($i = 1, 2, 3, \ldots, n$), suppose that the mean of the generated data is deliberately increased or decreased by $\Delta \mu$ for resampling of the observations with bootstrapping. As a result, high (low) values are likely to be resampled and low (high) values could be less likely to be selected. Thus, IBB can be obtained by allocating different weights $S_{i,n}$ depending on the following observation values (Equation (10)):

$$S_{i,n} = i/n. \tag{10}$$

The weight $S_{i,n}$ assigned after scaling and adjustment contributes to the probability of selection for the data observed in the IBB procedure. The average of the resampled data can be expressed as in Equation (11):

$$\tilde{\mu} = \frac{1}{\psi} \sum_{i=1}^{n} S_{i,n} x_i \tag{11}$$

where x_i represents the i-th incremental value and $\psi = \sum_{i=1}^{n} S_{i,n}$. The average amount of increase or decrease $\Delta\mu$ is shown in Equation (12):

$$\Delta\mu = \frac{1}{\psi}\sum_{i=1}^{n} S_{i,n} x_i - \frac{1}{n}\sum_{i=1}^{n} x_i. \qquad (12)$$

To obtain another value of $\Delta\mu$, the weights can be regeneralized in the order of the weight sequence (r); thus, $\Delta\widetilde{\mu}(r)$ is derived as follows:

$$\Delta\widetilde{\mu}(r) = \widetilde{\mu}(r) - \hat{\mu} = \frac{1}{\psi_r}\sum_{i=1}^{n} S_{i,n}^{r} x_i - \frac{1}{n}\sum_{i=1}^{n} x_i. \qquad (13)$$

If the average value of increase or decrease is given as $\Delta\mu$, then the weight "r" can be calculated accordingly. In this study, the selection of the weight sequence was performed using a Self-Organizing Migrating Algorithm with the objective function to minimize $[\Delta\mu - \Delta\widetilde{\mu}(r)]^2$. This approach follows a past study of extreme droughts during spring (March–May) in the Indochina peninsula [42]; however, no climate change indices for both the wet season (May–October) and the dry season (November–April) were integrated in that study, taking into account both the dipole mode in the tropical Indian Ocean and SST warming in the Pacific Ocean. In addition, the IBB technique was employed to generate resampled datasets for the IOD and the response to the intensity and frequency of rainfall in order to identify IOD-sensitive hotspots over the ICP. The statistical significance of the analysis results was assessed using the significance level of the 95th percentile.

3. Results

3.1. Seasonal Precipitation Patterns across the ICP

The ICP is a region in which monsoon rains occur in different seasons in association with seasonal winds and mountain areas. Geographically, the ICP has the Arakan Mountains in the west, the Bilauktung Mountains and the Dawna Mountains in the center, and the Annamese Mountains in the east. Meteorologically, the ICP is divided into three monsoon periods: the southwest monsoon during June–November, southeast monsoon during September–November, and northeast monsoon during November–February. This study considered the wet season (May–October) and the dry season (November–April) to identify the potential impact on regional rainfall associated with atmosphere–ocean feedback in the Indian and Pacific oceans.

Figure 4 shows the seasonal average precipitation during the wet and dry seasons across the ICP region during 1979–2018. The total precipitation during the wet season across the ICP is about 1000–1500 mm. In addition, it has been confirmed that precipitation variability is dependent on specific regions (Figure 4a). The precipitation variability was found to differ significantly between inland (<1000 mm) and coastal areas (>2000 mm). Precipitation on the western coast of Cambodia, the coast of western Thailand, and Myanmar during June–November is attributable to the influence of the southwest and southeast monsoons. Moreover, a clear difference in precipitation is evident between eastern and western parts of the Arakan Mountains in Myanmar. As water vapor from Bangorman decreases over the mountains, the Arakan Mountains show an arid climate to the east and a pattern of strong precipitation to the west.

During the dry season, total precipitation across the ICP is about 150–200 mm, indicating that rainfall variability is not significantly dependent on specific regions (Figure 4b). In particular, in the dry season, because of the influence of the northeast monsoon during November–February, high rainfall is received in central coastal areas of Vietnam, e.g., near the city of Danang. Similarly, in the case of Myanmar, eastern parts are dry because of the influence of the Arakan Mountains. The climatic characteristics of the ICP are distinctive not only because of the effects of monsoons and mountain areas, but also because of the characteristics of local areas and because of specific temporal effects. The precipitation patterns of the ICP are likely to change according to the characteristics of the wet and dry seasons, as well as because of the influence of ocean-related climate factors (e.g., the IOD and ENSO).

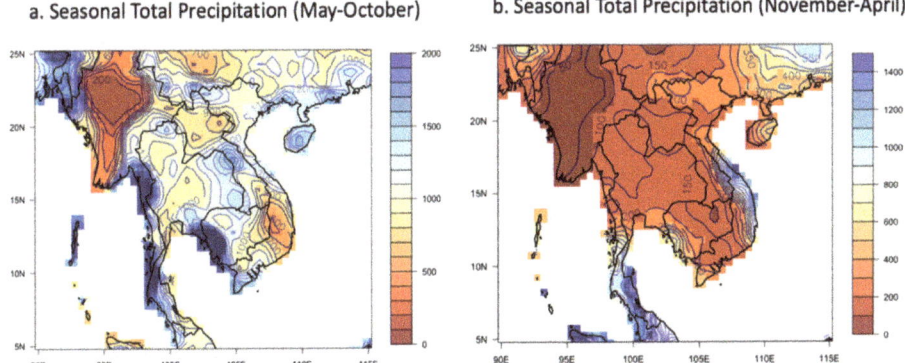

Figure 4. Average precipitation (mm) during the (**a**) wet and (**b**) dry seasons (1979–2018).

3.2. Spatiotemporal Variation in Precipitation over the ICP

Figures 5 and 6 illustrate the long-term trend of precipitation over the ICP during the period 1979–2018 for the wet and dry seasons, respectively. They show the results of the six major climate change indices that represent the magnitude and frequency of precipitation. For each figure, the direction of the trend is displayed in blue (increase) and red (decrease). Figure 5a,b and Figure 6a,b show the long-term trends of PRCPTOT and R95pTOT. These seasonal indices can be used to assess total precipitation. It can be seen that the characteristics of their spatial distribution are similar. During the wet season, there is a noticeable decrease in precipitation at the 5–10% significance level in northern Cambodia, some parts of Laos, and southern Thailand. In addition, it can be seen that there is a marked trend of increase at the 5–10% significance level in northwestern Myanmar, parts of western Thailand, central Vietnam, and southern parts of China (Figure 5a,b).

During the dry season, there is a noticeable increase in precipitation at the 5–10% significance level along eastern and southern coastal areas of the ICP (i.e., Vietnam and Cambodia) and some southern coastal regions of Thailand (Figure 6a,b). The R95pTOT climate index also shows a trend of increase in precipitation to the west of the Arakan Mountains in Myanmar (Figure 6b). Therefore, long-term changes in the pattern of precipitation across the ICP during the wet season show a trend of decrease (increase) in central inland areas (some coastal areas). During the dry season, there is a general trend of increase in precipitation across the ICP. Notably, the trend of increase in precipitation in southeastern coastal areas appears significant.

Figures 5c,d and 6c,d illustrate the long-term trends in RX1day and SDII. The RX1day and SDII climate indices can be used to assess rainfall intensity. It can be seen that the characteristics of the spatial distribution of the two indices are similar. Moreover, the characteristics of their spatial distribution are also similar to PRCPTOT and R95pTOT. It can be seen that during the rainy season the intensity of rainfall in central and northern Myanmar, central and southern Vietnam, and southern China increases, whereas the rainfall intensity decreases in Laos, Cambodia, northeastern Myanmar, and South Vietnam. During the dry season, rainfall intensity generally increases across the ICP, although it shows a clear pattern of decrease in Laos, as in the wet season.

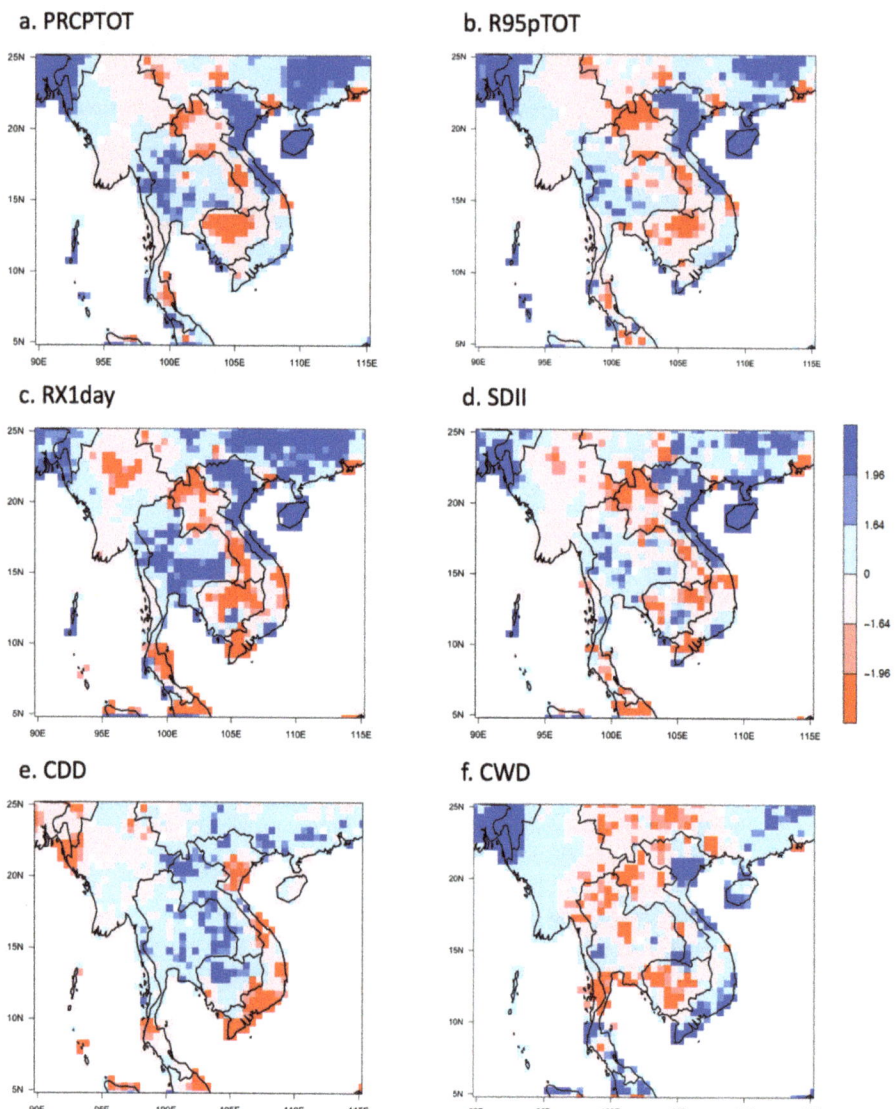

Figure 5. Long-term trend in seasonal precipitation for the wet season (May–October) over the ICP during 1979–2018. (**a–f**) show the analysis results of the six major climate change indices that reflect the magnitude and frequency of precipitation. In each panel, positive and negative trends are displayed in blue and red, respectively. The magnitude of Z is associated with the significance level, i.e., $|Z| > 1.64$ is for the 10% significance level and $|Z| > 1.96$ is for the 5% significance level.

Figures 5e,f and 6e,f show the long-term trends in CDD and CWD. The CDD and CWD indices can be used in assessment of droughts and floods, respectively. Therefore, it is unsurprising that the CDD and CWD indices exhibit opposite spatial distribution characteristics. During the rainy season, the CDD value across the ICP largely tends to increase, although it decreases in some coastal areas, e.g., Vietnam. The CWD index shows the reverse tendency.

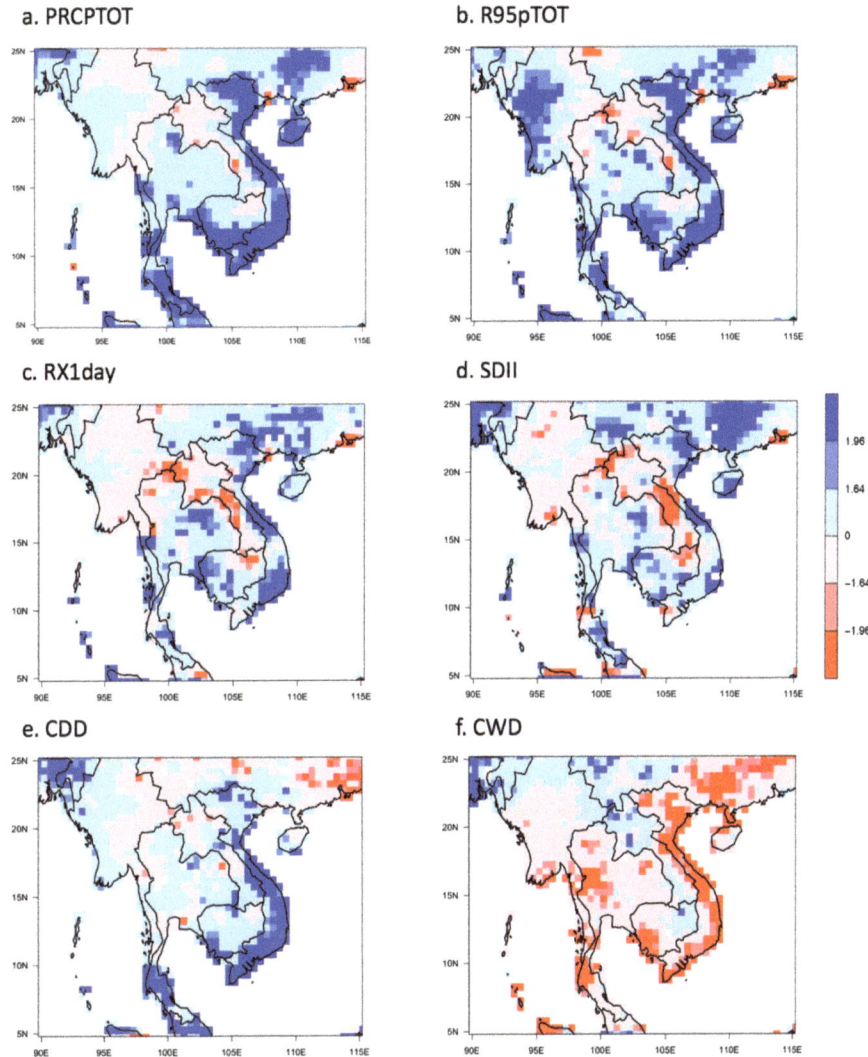

Figure 6. Long-term trend in seasonal precipitation for the dry season (November–April) over the ICP during 1979–2018. (**a–f**) show the analysis results of the six major climate change indices that reflect the magnitude and frequency of precipitation. In each panel, positive and negative trends are displayed in blue and red, respectively. The magnitude of Z is associated with the significance level, i.e., |Z| > 1.64 is for the 10% significance level and |Z| > 1.96 is for the 5% significance level.

During the dry season, an increase (decrease) of the CDD (CWD) index can be clearly observed at the 5–10% significance level (Figure 6e,f). The CDD index increases along the southeast coast of the ICP, e.g., in areas of Vietnam, Cambodia, and southern Thailand, whereas the CWD index exhibits the opposite trend. An increase (decrease) in the CDD index suggests that drought is more (less) likely to occur, while a decrease (increase) in the CWD index means that the occurrence of drought is less (more) likely. Therefore, during the rainy season, floods are expected to increase along the southeastern coast of the ICP (e.g., in Vietnam, Cambodia, and Thailand), while drought is more likely to occur during the dry season.

3.3. Precipitation Variability Associated With the IOD and ENSO

The IOD, Asian monsoon, and other regional climatological patterns can lead to local or global climate change, particularly in Indian Ocean Rim countries, which can cause severe flooding or droughts depending on IOD variability [26]. Composite analysis can clarify the role of the Southeast Asian Summer Monsoon in precipitation variability across the ICP region associated with years of strong IOD and ENSO after identifying that tropical climate phenomena are the main factors that influence precipitation variability over the ICP during the wet and dry seasons. However, this role differs depending on the combination of the two climate phenomena and on the season.

Figure 7 shows the results of composite rainfall anomalies (shown as a percentage relative to normal) over the ICP during the wet and dry seasons in relation to the IOD and ENSO. The patterns of rainfall anomalies indicate significant difference between positive and negative IOD years. For positive IOD years, the wet season rainfall (Figure 7a) shows a decrease of <20% in southern parts of the ICP, whereas there is a marked increase in rainfall centered over the Arakan Mountains in western Myanmar. It can be seen that the amount of rainfall received during the dry season (Figure 7c) is similar to that in the wet season, but there is 40–50% less rainfall than usual in certain mainland regions of Southeast Asia, especially Yangon and Mawlamyine in Myanmar and in eastern Cambodia.

In negative IOD years, intense positive anomalies of rainfall can be seen in central Cambodia and southern parts of Vietnam. A slight strong-pitched anomaly pattern is evident during the wet season (Figure 7b) around the coastline of both Bangladesh and Myanmar, whereas weak-pitched positive anomalies (about 10–15% relative to the long-term average) are found throughout the ICP. However, changes in rainfall pattern are not evident during the dry season (Figure 7d), and although the amount varies depending on region, rainfall is generally >30–50% above the long-term average. As in the wet season, the dry season also shows relatively strong positive rainfall patterns with positive anomalies of >80–100% in Cambodia and both central and southern Vietnam.

Sometimes droughts and flooding are likely to converge because of remote connections during IOD–ENSO periods, and they can have a significant impact on the modulation of the large-scale oceanic and atmospheric environment, especially in the Indian Ocean and in Pacific Rim countries [25,26,43]. Thus, consideration of both combined and independent effects of ENSO and the IOD on seasonal precipitation variability can provide improved predictive expertise, and reveal new insight into tropical climate change and global warming impacts [28].

Figure 8 shows composite rainfall anomalies (November–April) during positive and negative IOD years that coincided with ENSO. During positive IOD and El Niño years (Figure 8a), there is less rainfall than usual across Thailand, Cambodia, southern Laos, and Vietnam. In particular, southern regions of Myanmar (from Yangon to Mawlamyine) that border the Andaman Sea show a distinct decrease in rainfall by more than 50% in comparison with the long-term mean (1981–2010). However, in contrast, there is 20–40% more rainfall than usual in northern parts of the ICP, e.g., northern Myanmar (around 97E, 22N), northeastern parts of Laos (around 102E, 21N), and Vietnam. Furthermore, in Guangzhou in China, rainfall is up to 60% higher in comparison with average years. These rainfall signals are stronger in WP El Niño years than in CT El Niño years (figures not shown). During negative IOD and La Niña years (Figure 8b), rainfall above the long-term average is observed throughout the ICP, except for parts of central Laos (around 105E, 17N) and northern Vietnam (around 107E, 21N). The pattern of increased rainfall appears strongly throughout Myanmar and regions around Ho Chi Minh City in Vietnam. However, in the region adjacent to India and Bangladesh, as well as the Shenzhen area of China, strong negative anomalies are evident.

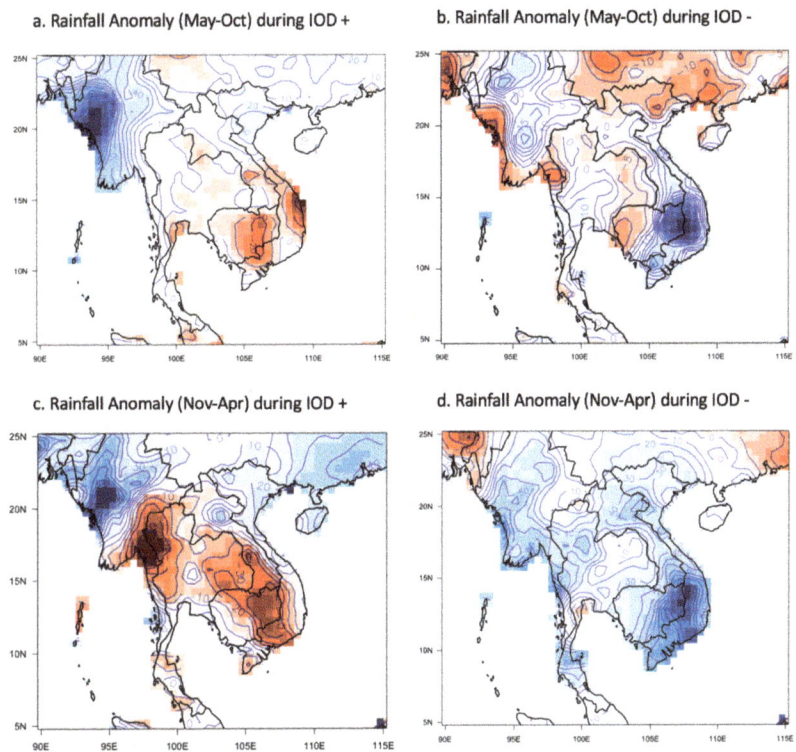

Figure 7. Composite of seasonal rainfall anomaly (%) during positive and negative IOD years: (**a**) rainfall anomaly in wet season during positive IOD years, (**b**) rainfall anomaly in wet season during negative IOD years, (**c**) rainfall anomaly in dry season during positive IOD years, and (**d**) rainfall anomaly in dry season during negative IOD years. Positive (negative) values show increasing (decreasing) rainfall departure from the long-term average (1981–2010).

3.4. Identification of IOD-Sensitive Hotspots through IBB Simulations

Section 3.3 discussed the significant impact on rainfall anomalies in the ICP that are attributable to the combined or independent effects of ENSO and the IOD. In particular, both positive IOD events with El Niño and negative IOD events with La Niña interact in modulating rainfall anomalies over the ICP. The IOD and ENSO are strongly correlated, and their variations are mutually forced or triggered [43,44]. For the period of 1979–2018, the correlation between the peak phase of the IOD and the two types of El Niño index proposed by Ren and Jin [36] was analyzed. The results showed that the IOD has a strong positive correlation with the CT El Niño (N_{CT}) (ρ = 0.4850, p-value = 0.0018). However, the IOD also has positive correlation with the WP El Niño (N_{WP}), but not at a statistically significant level (ρ = 0.110, p-value = 0.5013). These results are also reflected in the results of the IBB simulation (Figure 9). Figure 9 shows the results of 1000 simulations for the N_{CT} and N_{WP} indices performed by applying the IBB technique to the IOD index based on historical observations for the period of 1979–2017. By applying a +1 SD increase of the IOD, the mean difference between the observation of N_{CT} and simulated N_{CT} shows a statistically significant increase at the 95% significance level (diff. = 0.392, Interquartile range (IQR) = 0.228). However, the difference in the mean value of the N_{WP} index, although increased slightly, is not statistically significant (diff. = 0.097, IQR = 0.094). By applying a −1 SD decrease of the IOD, the simulation results show changes similar to the case with a +1 SD increase of the IOD (N_{CT}: diff. = 0.360,

IQR = 0.108, N_{WP}: diff. = 0.088, IQR = 0.098). Therefore, for changes in the IOD, the linear increase (or decrease) in the N_{CT} index is more pronounced than the change in the N_{WP} index.

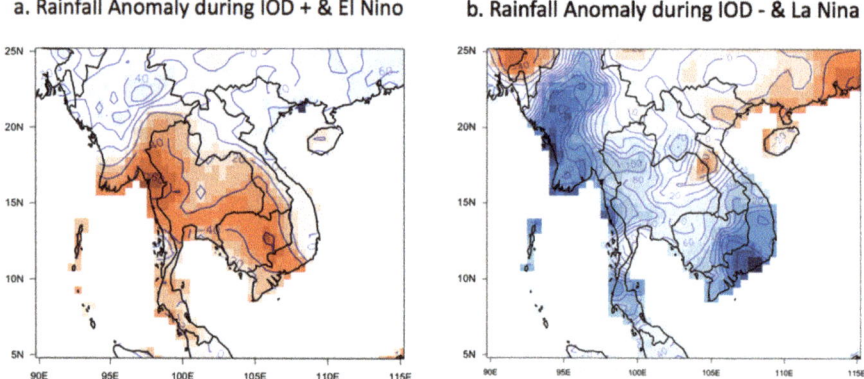

Figure 8. Composite rainfall anomaly in dry season (November–April) associated with the IOD and El Niño–Southern Oscillation (ENSO): (**a**) rainfall anomaly during years with positive IOD and El Niño, and (**b**) rainfall anomaly during years with negative IOD and La Niña. Positive (negative) values show increasing (decreasing) rainfall departure from the long-term average (1981–2010).

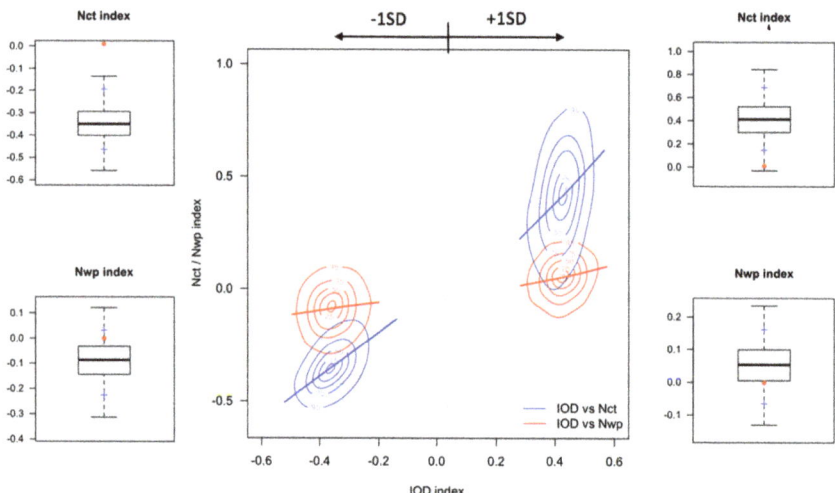

Figure 9. Mean differences of the two types of El Niño with ±1 SD of the IOD. In the main panel, contours (5th, lower quadrant, median, upper volatile, and 95th level) summarize the IOD index and CT El Niño (N_{CT}) or WP El Niño (N_{WP}) index using the intentionally biased bootstrapping model. Both left and right panels deliberately apply ±1 SD of the IOD to show results of 1000 simulations for the N_{CT} and N_{WP} indices. Red dots in each panel represent the average value of the observations.

The spatiotemporal connection between SST and winds shows strong coupling through precipitation and ocean dynamics [24]. This dipole mode accounts for about 12% of SST variability in the Indian Ocean, and its duration of activity can greatly affect both the intensity and the frequency of rainfall in the Indian Ocean Rim countries, including the ICP. Based on statistical simulations of historical observations (1979–2018), Figures 10 and 11 show rainfall variation and the most sensitive hotspot areas in the wet and dry seasons of the ICP attributable to IOD changes.

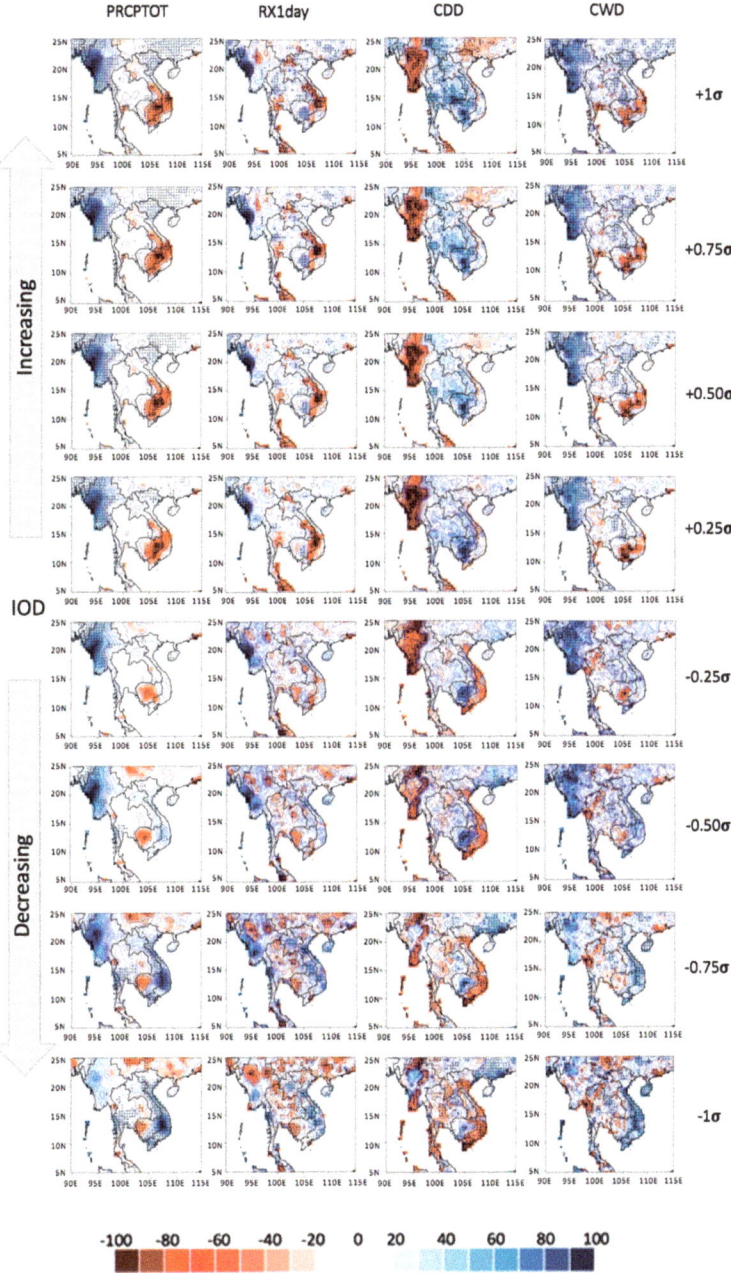

Figure 10. Spatial distributions of the percentage changes in major precipitation indices for the wet season (May–October) over the ICP region for intentional increases or decreases by 0.25 SD of the IOD index using the intentionally biased bootstrapping simulation. For each panel, the statistically significant area of change at the 95% significance level is shown by an "x" symbol.

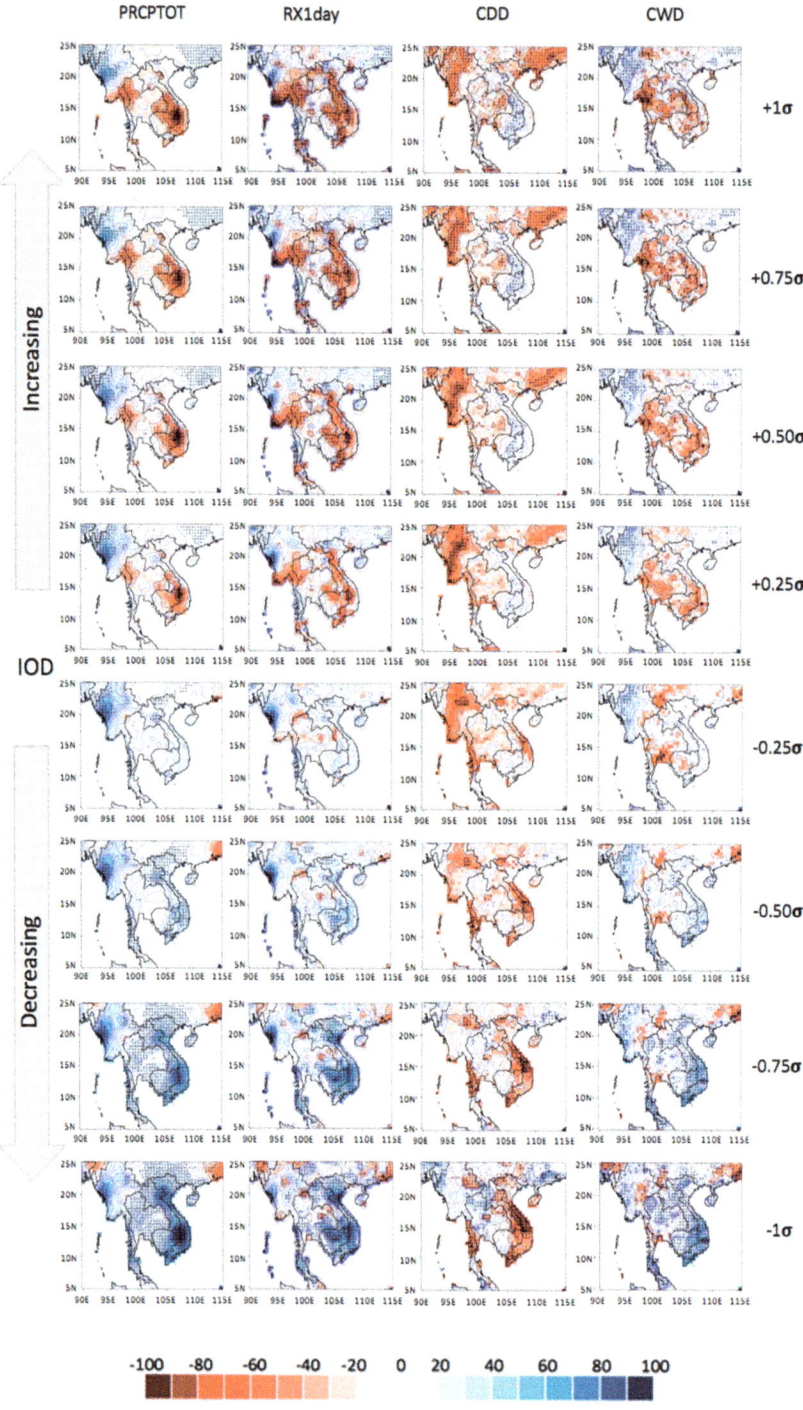

Figure 11. Same as Figure 10, but for the dry season (November–April) over the ICP region.

The spatial distribution of differences in PRCPTOT is shown in Figure 10, given the condition of increases or decreases by 0.25 SD of the IOD index in the wet season. For a +1 SD increase of the IOD, PRCPTOT is >90% higher than usual throughout Myanmar, and weak positive anomaly patterns are evident in southwestern China. In contrast, a pattern of decrease of PRCPTOT of 15–20% less than the long-term average is evident in Cambodia and southern Vietnam, i.e., in areas of the downstream reaches of the Mekong River. However, no statistically significant changes occur in the central ICP region, except in some parts of central Laos and Thailand. This spatial distribution of rainfall anomaly is also found for the RX1day index, although occasional patterns of increase or decrease are evident and the spatial extent is reduced. In addition, throughout Myanmar, the CDD index is decreased by >25% in comparison with the long-term average year, while the CWD index is increased by 35–50%. For the CDD index, a statistically significant pattern of decrease is found across Vietnam, Cambodia, and Laos. The most significant changes in the CWD index are across Myanmar (increase of 35–50%), southern Cambodia, and the southeast coast of Vietnam (decrease of 15–20%). The other ICP regions generally show a pattern of weak increase in terms of CWD. For a −1 SD decrease of the IOD, PRCPTOT, RX1day, and CWD all show distinct patterns of increase in the Laos and Vietnam basins, while the CDD index shows a predominant pattern of decrease, except in certain areas. Analysis indicates that other regions have a reverse pattern compared with the case of the +1 SD increase of the IOD. Consequently, it is determined that changes in rainfall during the wet season in the ICP region are sensitive to changes in the IOD.

Given the condition of increases or decreases by 0.25 SD of the IOD index for the dry season, the spatial distribution of the rainfall indices is shown in Figure 11. For a decrease of −1 SD of the IOD, there is more rainfall (PRCPTOT and RX1day) than usual throughout the ICP, especially in Laos and Vietnam. For a +1 SD increase of the IOD, negative anomaly patterns of PRCPTOT are dominant in southern Vietnam, eastern Cambodia, and northeastern Thailand, while weak patterns of positive anomaly are evident in areas of Myanmar and South China. Compared with the changes in the rainfall indices during the wet season, changes in the rainfall indices are intensified and the spatial influence is more extensive. However, for the CDD and CWD indices, either the positive anomaly patterns are weakened or the negative anomaly patterns appear for a +1 SD increase of the IOD. Especially for the CWD index, a pattern of decrease by more than 10–20% compared with the long-term average is found in Thailand, whereas the Myanmar region shows a pattern of increase of 15–25%. In this study, we simulated the changes in both wet and dry season rainfall across the ICP according to intentional IOD changes, and IOD-sensitive hotspots were verified through quantitative analysis. The findings of this study could help elucidate the long-term changes in rainfall expected in the ICP region in a changing climate.

4. Summary and Conclusions

This study analyzed changes in the magnitude and frequency of precipitation during the dry and wet seasons over the ICP, taking into account both the dipole mode in the tropical Indian Ocean and SST warming in the Pacific Ocean. The main results are summarized as follows.

1. According to analyses of the long-term trend in seasonal rainfall across the ICP during 1979–2018, rainfall showed significant decreases in northern Cambodia, parts of Laos, and southern Thailand during the wet season (May–October). Moreover, significant increases were evident in northwestern Myanmar, some parts of western Thailand, central Vietnam, and southern China. During the dry season (November–April), PRCPTOT rose noticeably in eastern and southern coastal areas of the ICP (i.e., Vietnam and Cambodia) and some southern coastal regions of Thailand.

2. During the wet season, the CDD index increased or decreased in some coastal areas such as Vietnam. However, during the dry season, increases in CDD and decreases in CWD were evident in the ICP. In particular, a pattern of decline in CWD dominated southeastern coastal areas of the ICP, including Vietnam, Cambodia, and southern Thailand.

3. The IOD showed strong positive correlation with the CT El Niño. However, although the IOD exhibited positive correlation with the WP El Niño, the relationship was not statistically significant. The variability of rainfall pattern and amount across the ICP was confirmed to be amplified by combined and independent effects of ENSO and the IOD. During years of positive IOD and El Niño, there was less rainfall than usual throughout Thailand and Cambodia, southern Laos, and Vietnam. In particular, the southern part of Myanmar, which borders the Andaman Sea, showed a decrease in regional rainfall of >50% in comparison with the long-term average. In contrast, northern parts of India and China, including Myanmar, northeastern Laos, and Vietnam, received 20–40% more rainfall than usual. Years with a negative IOD mode and La Niña showed rainfall above the long-term average across the ICP, except for certain parts, e.g., Central Laos and northern Vietnam.

4. Through application of the IBB technique, this study simulated the change of rainfall pattern and amount across the ICP for the wet and dry seasons according to intentional IOD changes, and IOD-sensitive hotspots were verified through quantitative analysis. For the wet season, a +1 SD increase of the IOD resulted in >90% more PRCPTOT than usual across Myanmar in the northwestern ICP. Conversely, in Cambodia and southern Vietnam, rainfall patterns were 15–20% less than the long-term average in the region of the lower Mekong River. In addition, the CDD index decreased throughout Myanmar by >25% compared with the long-term average. The most significant change in the CWD index was in Myanmar, i.e., a 35–50% increase. However, a pattern of decrease appeared across the southeastern coast of the ICP in southern Cambodia and Vietnam. For a +1 SD increase of the IOD in the dry season, negative anomaly patterns of PRCPTOT were found to be dominant in South Vietnam, eastern Cambodia, and northeastern Thailand, and more rainfall than usual occurred throughout the ICP, especially in Laos and Vietnam, when considering a −1 SD decrease of the IOD.

Although the results of this study are based on limited observations, they provide a clear perspective on the sensitivity of local precipitation to atmosphere–ocean interactions, and they reveal the potential future impact of statistical changes to the IOD, improving our understanding of the associated regional impact on precipitation under the effects of climate change.

Author Contributions: Conceptualization, Resources, Formal analysis, Writing—original draft, J.-S.K. and P.X.; Conceptualization, Methodology, Writing—review & editing, S.-K.Y. and T.L.; Writing—review & editing, L.X. All authors have read and agreed to the published version of the manuscript.

Funding: This research is supported by the National Natural Science Foundation of China (NSFC Grant Nos. 41890822 and 51525902). This work was also supported by the National Research Foundation of Korea (NRF) grant funded by the Korean Government (MEST) (2018R1A2B6001799). We also appreciate the support of the State Key Laboratory of Water Resources and Hydropower Engineering Science, Wuhan University.

Conflicts of Interest: The authors declare no conflict of interest.

References

1. Allan, R.P.; Soden, B.J. Atmospheric warming and the amplification of precipitation extremes. *Science* **2008**, *321*, 1481–1484. [CrossRef] [PubMed]
2. Kim, J.S.; Jain, S. Precipitation trends over the Korean peninsula: Typhoon-induced changes and a typology for characterizing climate-related risk. *Environ. Res. Lett.* **2011**, *6*, 034033. [CrossRef]
3. Ge, F.; Zhi, X.; Babar, Z.A.; Tang, W.; Chen, P. Interannual variability of summer monsoon precipitation over the Indochina Peninsula in association with ENSO. *Theor. Appl. Climatol.* **2017**, *128*, 523–531. [CrossRef]
4. Kang, H.Y.; Kim, J.S.; Kim, S.Y.; Moon, Y.I. Changes in High-and Low-Flow Regimes: A Diagnostic Analysis of Tropical Cyclones in the Western North Pacific. *Water Resour. Manag.* **2017**, *31*, 3939–3951. [CrossRef]
5. Kim, J.S.; Son, C.Y.; Moon, Y.I.; Lee, J.H. Seasonal rainfall variability in Korea within the context of different evolution patterns of the central Pacific El Niño. *J. Water Clim. Chang.* **2017**, *8*, 412–422. [CrossRef]
6. Gao, Q.; Kim, J.S.; Chen, J.; Chen, H.; Lee, J.H. Atmospheric Teleconnection-Based Extreme Drought Prediction in the Core Drought Region in China. *Water* **2019**, *11*, 232. [CrossRef]

7. Chi, X.; Yin, Z.; Wang, X.; Sun, Y. Spatiotemporal variations of precipitation extremes of China during the past 50 years (1960–2009). *Theor. Appl. Climatol.* **2016**, *124*, 555–564. [CrossRef]
8. Gu, X.; Zhang, Q.; Singh, V.P.; Shi, P. Changes in magnitude and frequency of heavy precipitation across China and its potential links to summer temperature. *J. Hydrol.* **2017**, *547*, 718–731. [CrossRef]
9. Choi, J.H.; Yoon, T.H.; Kim, J.S.; Moon, Y.I. Dam rehabilitation assessment using the Delphi-AHP method for adapting to climate change. *J. Water Resour. Plan. Manag.* **2018**, *144*, 06017007. [CrossRef]
10. Croitoru, A.E.; Chiotoroiu, B.C.; Todorova, V.I.; Torică, V. Changes in precipitation extremes on the Black Sea Western Coast. *Glob. Planet. Chang.* **2013**, *102*, 10–19. [CrossRef]
11. IPCC: Climate change 2013: The Physical Science Basis. *Contribution of Working Group I to the Fifth Assessment Report of the Intergovernmental Panel on Climate Change*; Cambridge University Press: Cambridge, UK, 2013.
12. Hirsch, R.M.; Archfield, S.A. Flood trends: Not higher but more often. *Nat. Clim. Chang.* **2015**, *5*, 198. [CrossRef]
13. Donat, M.G.; Lowry, A.L.; Alexander, L.V.; O'Gorman, P.A.; Maher, N. More extreme precipitation in the world's dry and wet regions. *Nat. Clim. Chang.* **2016**, *6*, 508–514. [CrossRef]
14. Mirza, M.M.Q. Climate change and extreme weather events: Can developing countries adapt? *Clim. Policy* **2003**, *3*, 233–248. [CrossRef]
15. Yin, J.; Yin, Z.E.; Zhong, H.D.; Xu, S.Y.; Hu, X.M.; Wang, J.; Wu, J.P. Monitoring urban expansion and land use/land cover changes of Shanghai metropolitan area during the transitional economy (1979–2009) in China. *Environ. Monit. Assess.* **2011**, *177*, 609–621. [CrossRef] [PubMed]
16. Chen, D.; Cane, M.A. El Niño prediction and predictability. *J. Comput. Phys.* **2008**, *227*, 3625–3640. [CrossRef]
17. Ashok, K.; Yamagata, T. The El Niño with a difference. *Nature* **2009**, *461*, 481–484. [CrossRef]
18. Pradhan, P.K.; Preethi, B.; Ashok, K.; Krishna, R.; Sahai, A.K. Modoki, Indian Ocean Dipole, and western North Pacific typhoons: Possible implications for extreme events. *J. Geophys. Res.* **2011**, *116*, D18108. [CrossRef]
19. Kim, J.S.; Zhou, W.; Wang, X.; Jain, S. El Nino Modoki and the summer precipitation variability over South Korea: A diagnostic study. *J. Meteorol. Soc. Jpn.* **2012**, *90*, 673–684. [CrossRef]
20. Yoon, S.K.; Kim, J.S.; Lee, J.H.; Moon, Y.I. Hydrometeorological variability in the Korean Han River Basin and its sub-watersheds during different El Niño phases. *Stoch. Environ. Res. Risk Assess.* **2013**, *27*, 1465–1477. [CrossRef]
21. Son, C.Y.; Kim, J.S.; Moon, Y.I.; Lee, J.H. Characteristics of tropical cyclone-induced precipitation over the Korean River basins according to three evolution patterns of the Central-Pacific El Niño. *Stoch. Environ. Res. Risk Assess.* **2014**, *28*, 1147–1156. [CrossRef]
22. Wang, X.; Zhou, W.; Li, C.Y.; Wang, D.X. Comparison of the impact of two types of El Niño on tropical cyclone genesis over the South China Sea. *Int. J. Climatol.* **2014**, *34*, 2651–2660. [CrossRef]
23. Piechota, T.C.; Chiew, F.H.S.; Dracup, J.A.; McMachon, T.A. Seasonal streamflow forecasting in eastern Australia and the El Niño-Southern Oscillation. *Water Res. Res.* **1998**, *34*, 3035–3044. [CrossRef]
24. Saji, N.H.; Goswami, B.N.; Vinayachandran, P.N.; Yamagata, T. A Dipole Mode in the tropical Indian Ocean. *Nature* **1999**, *401*, 360363. [CrossRef] [PubMed]
25. Mahala, B.K.; Nayak, B.K.; Mohanty, P.K. Impacts of ENSO and IOD on tropical cyclone activity in the Bay of Bengal. *Nat. Hazards* **2015**, *75*, 1105–1125. [CrossRef]
26. Iqbal, A.; Hassan, S.A. ENSO and IOD analysis on the occurrence of floods in Pakistan. *Nat. Hazards* **2018**, *91*, 879–890. [CrossRef]
27. Webster, P.J.; Moore, A.M.; Loschnigg, J.P.; Leben, R.R. Coupled ocean–atmosphere dynamics in the Indian Ocean during 1997–1998. *Nature* **1999**, *401*, 356–360. [CrossRef] [PubMed]
28. Ashok, K.; Guan, Z.; Yamagata, T. Impact of the Indian Ocean Dipole on the relationship between the Indian monsoon rainfall and ENSO. *Geophys. Res. Lett.* **2001**, *28*, 4499–4502. [CrossRef]
29. Ashok, K.; Guan, Z.; Yamagata, T. Influence of the Indian Ocean dipole on the Australian winter rainfall. *Geophys. Res. Lett.* **2003**, *30*, 1821. [CrossRef]
30. McPhaden, M.J.; Zebiak, S.E.; Glantz, M.H. ENSO as an integrating concept in Earth Science. *Science* **2006**, *314*, 1740–1745. [CrossRef]
31. Kosaka, Y.; Xie, S.P. Recent global-warming hiatus tied to equatorial Pacific surface cooling. *Nature* **2013**, *501*, 403. [CrossRef]

32. Lee, S.K.; Park, W.; Baringer, M.O.; Gordon, A.L.; Huber, B.; Liu, Y. Pacific origin of the abrupt increase in Indian Ocean heat content during the warming hiatus. *Nat. Geosci.* **2015**, *8*, 445. [CrossRef]
33. Liu, W.; Xie, S.P.; Lu, J. Tracking ocean heat uptake during the surface warming hiatus. *Nat. Commun.* **2016**, *7*, 10926. [CrossRef] [PubMed]
34. Schneider, U.; Ziese, M.; Meyer-Christoffer, A.; Finger, P.; Rustemeier, E.; Becker, A. The new portfolio of global precipitation data products of the Global Precipitation Climatology Centre suitable to assess and quantify the global water cycle and resources. *Proc. IAHS* **2016**, *374*, 29–34. [CrossRef]
35. Karl, T.R.; Nicholls, N.; Ghazi, A. Clivar/GCOS/WMO workshop on indices and indicators for climate extremes workshop summary. In *Weather and Climate Extremes*; Springer: Berlin, Germany, 1999; pp. 3–7.
36. Ren, H.L.; Jin, F.F. Nino indices for two types of ENSO. *Geophys. Res. Lett.* **2011**, *38*, L04704. [CrossRef]
37. Mann, H.B. Nonparametric Tests Against Trend. *Econometrica* **1945**, *13*, 245–259. [CrossRef]
38. Kendall, M.G.; Gibbons, J.D. *Rank Correlation Methods*; Arnold, E., Ed.; Springer: Berlin, Germany, 1990.
39. Hamed, K.H.; Rao, A.R. A modified Mann-Kendall trend test for autocorrelated data. *J. Hydrol.* **1998**, *204*, 182–196. [CrossRef]
40. Davison, A.C.; Hinkley, D.V.; Young, G.A. Recent developments in bootstrap methodology. *Stat. Sci.* **2003**, *18*, 141–157. [CrossRef]
41. Lee, T. Climate change inspector with intentionally biased bootstrapping (CCIIBB ver. 1.0)–methodology development. *Geosci. Model Dev.* **2017**, *10*, 525–536. [CrossRef]
42. Gao, Q.G.; Sombutmounvong, V.; Xiong, L.; Lee, J.H.; Kim, J.S. Analysis of Drought-Sensitive Areas and Evolution Patterns through Statistical Simulations of the Indian Ocean Dipole Mode. *Water* **2019**, *11*, 1302. [CrossRef]
43. Lestari, R.K.; Koh, T.Y. Statistical evidence for asymmetry in ENSO–IOD interactions. *Atmos. Ocean* **2016**, *54*, 498–504. [CrossRef]
44. Yuan, Y.; Li, C. Decadal variability of the IOD-ENSO relationship. *Chin. Sci. Bull.* **2008**, *53*, 1745–1752. [CrossRef]

© 2020 by the authors. Licensee MDPI, Basel, Switzerland. This article is an open access article distributed under the terms and conditions of the Creative Commons Attribution (CC BY) license (http://creativecommons.org/licenses/by/4.0/).

Article

Evaluation of Multi-Satellite Precipitation Datasets and Their Error Propagation in Hydrological Modeling in a Monsoon-Prone Region

Jie Chen [1,2,*], Ziyi Li [1], Lu Li [3], Jialing Wang [1], Wenyan Qi [1], Chong-Yu Xu [4] and Jong-Suk Kim [1]

1. State Key Laboratory of Water Resources and Hydropower Engineering Science, Wuhan University, Wuhan 430072, China; 2013301580011@whu.edu.cn (Z.L.); 2015202060025@whu.edu.cn (J.W.); wenyan.qi@whu.edu.cn (W.Q.); jongsuk@whu.edu.cn (J.-S.K.)
2. Hubei Key Laboratory of Water System Science for Sponge City Construction, Wuhan University, Wuhan 430072, China
3. NORCE Norwegian Research Centre, Bjerknes Centre for Climate Research, Jahnebakken 5, NO-5007 Bergen, Norway; luli@norceresearch.no
4. Department of Geosciences, University of Oslo, Sem Saelands vei 1, P.O. Box 1047, Blindern, NO-0316 Oslo, Norway; c.y.xu@geo.uio.no
* Correspondence: jiechen@whu.edu.cn

Received: 13 September 2020; Accepted: 26 October 2020; Published: 30 October 2020

Abstract: This study comprehensively evaluates eight satellite-based precipitation datasets in streamflow simulations on a monsoon-climate watershed in China. Two mutually independent datasets—one dense-gauge and one gauge-interpolated dataset—are used as references because commonly used gauge-interpolated datasets may be biased and unable to reflect the real performance of satellite-based precipitation due to sparse networks. The dense-gauge dataset includes a substantial number of gauges, which can better represent the spatial variability of precipitation. Eight satellite-based precipitation datasets include two raw satellite datasets, Precipitation Estimation from Remotely Sensed Information using Artificial Neural Networks (PERSIANN) and Climate Prediction Center MORPHing raw satellite dataset (CMORPH RAW); four satellite-gauge datasets, Tropical Rainfall Measuring Mission 3B42 (TRMM), PERSIANN Climate Data Record (PERSIANN CDR), CMORPH bias-corrected (CMORPH CRT), and gauge blended datasets (CMORPH BLD); and two satellite-reanalysis-gauge datasets, Multi-Source Weighted-Ensemble Precipitation (MSWEP) and Climate Hazards Group InfraRed Precipitation with Stations (CHIRPS). The uncertainty related to hydrologic model physics is investigated using two different hydrological models. A set of statistical indices is utilized to comprehensively evaluate the precipitation datasets from different perspectives, including detection, systematic, random errors, and precision for simulating extreme precipitation. Results show that CMORPH BLD and MSWEP generally perform better than other datasets. In terms of hydrological simulations, all satellite-based datasets show significant dampening effects for the random error during the transformation process from precipitation to runoff; however, these effects cannot hold for the systematic error. Even though different hydrological models indeed introduce uncertainties to the simulated hydrological processes, the relative hydrological performance of the satellite-based datasets is consistent in both models. Namely, CMORPH BLD performs the best, which is followed by MSWEP, CMORPH CRT, and TRMM. PERSIANN CDR and CHIRPS perform moderately well, and two raw satellite datasets are not recommended as proxies of gauged observations for their worse performances.

Keywords: satellite-based precipitation; hydrological modeling; error propagation; monsoon-climate watershed

1. Introduction

Precipitation is one of the most important meteorological variables in the hydrologic cycle and is often used as the fundamental input to environmental models for agricultural, meteorological, and hydrological studies [1]. However, precipitation measured by pluviometers usually suffers from many problems, such as sparse station distribution at high altitudes or in rural areas, missing data, and short time periods [2]. Meanwhile, artificial errors in measurements are inevitable [3]. In addition, surface observational networks have indicated decreasing coverages and spatial densities, which may limit the future capacity to measure precipitation for many parts of the world [4,5].

As a proxy for gauged precipitation, gridded precipitation with high spatial and temporal coverage has been developed, which can be generally classified into three categories based on different data sources: (1) gauge-interpolated, (2) reanalysis-based, and (3) satellite-based precipitation [6–10].

Gauge-interpolated precipitation, such as the Global Precipitation Climatology Centre (GPCC) [11] and Climate Research Unit (CRU) [12], is generated by interpolating gauged data to grids with different spatial resolutions [13,14]. Thiessen polygons, Kriging, and inverse distance weighting (IDW) are the most widely used interpolation algorithms [2,15]. More sophisticated interpolation methods take extra geographical or physical information into consideration, such as topography and atmospheric lapse rate [13,14].

Reanalysis-based precipitation is produced by assimilating various observations (e.g., weather stations, satellites, ships, and buoys) into a climate model to generate various meteorological variables with a consistent spatial and temporal resolution [4,16–18]. The reliability of reanalysis-based precipitation relies on assimilated observations, climate model parameters, and the interactions between models and observations. Several reanalysis datasets have been made freely available, such as the National Centers for Environmental Prediction/National Center for Atmosphere Research Reanalysis (NCEP/NCAR) [19], the European Centre for Medium-Range Weather Forecasts Reanalysis (ERA) [20], and the NCEP Climate Forest System Reanalysis (CFSR) [21].

Satellite-based precipitation, with a global and continuous temporal scale, estimates precipitation using polar-orbiting passive microwave (PMW) sensors on low-Earth-orbiting satellites and geosynchronous infrared (IR) sensors on geostationary satellites [22–24]. PMW sensors could observe the emissions and lower-atmosphere scattering signals of rainfall, snow, and ice contents, while IR sensors indirectly measure the lower-level rainfall rate by collecting cloud-top temperature and cloud height [25]. Usually, in regions with gauges, satellite-based datasets are modified by gauged measurements to offset their limited abilities [26]. Over the past 30 years, a number of precipitation datasets that combine gauges, PMW, and IR data to produce precipitation estimates are available with the spatial resolution on 0.25 latitude/longitude or finer. These include monthly Global Precipitation Climatology Project (GPCP) [27], daily Precipitation Estimation from Remotely Sensed Information Using Artificial Neural Networks (PERSIANN) [28], Tropical Rainfall Measuring Mission (TRMM) Multi-satellite Precipitation Analysis (TMPA) [29], and Global Precipitation Measurement (GPM) [30].

Although gauge-interpolated and reanalysis-based precipitation may be more appropriate for climate change studies for their long-term data records, it is often difficult to verify their reliabilities in regions with sparse weather station networks [31,32]. Satellite-based datasets could estimate precipitation with a global and homogeneous spatial coverage; this spatial continuity could provide valuable information for hydrological modeling, especially for ungauged watersheds. In recent years, a few global-scale studies are revealing that the performances of satellite-based datasets differ regionally and temporally and correlate to topography, seasonality, and climatology [6,23,33,34]. However, the lack of global and sufficiently dense precipitation references makes these results of satellite-based precipitation still unreliable and inadequate for operational purposes, such as flood forecasting [35]. Therefore, a regional ground validation of satellite-based precipitation datasets based on dense gauges references, especially for their hydrological performances, still requires to be conducted [36–44]. Although most studies revealed the potential of satellite-based precipitation datasets for hydrological simulations, they also report error sources during the hydrological modeling of satellite-based datasets. Generally, two main sources are (1) the error of the satellite-based datasets and (2) the error propagation of satellite-based datasets through the hydrological model [45].

The monsoon regions, having an obvious seasonal variation of precipitation, have always been a research focus of satellite-based precipitation datasets [46–52]. For example, Prakash et al. [51] compared four satellite-based precipitation datasets (Climate Prediction Center MORPHing-raw satellite dataset (CMORPH RAW), Naval Research Laboratory (NRL)-blended, PERSIANN, and TRMM 3B42) with the gauged-interpolated dataset in one Indian monsoon region with respect to their abilities to simulate the seasonal rainfall and the rainfall detection abilities over regions with diverse topography. The results show that although all four datasets underestimate the summer seasonal mean rainfall (June to September), TRMM 3B42 generally performs better than the other three datasets mainly due to its incorporation of rain gauge observations. Mou et al. [49] compared five satellite-based precipitation datasets (TRMM 3B42, its real-time dataset TRMM 3B42RT, GPCP-1DD, PERSIANN Climate Data Record (PERSIANN CDR), and CMORPH RAW) and a gauge-interpolated dataset (Asian Precipitation-Highly Resolved Observational Data Integration Towards Evaluation of Water Resources (APHRODITE)) at daily, monthly, seasonal, and annual scales with rain gauges over Malaysia. It was found that TRMM 3B42 and APHRODITE performed the best, while PERSIANN CDR slightly overestimated observed precipitation, and the other three satellite-based datasets showed the worst performance. In addition, all six precipitation datasets show better performances in southern Peninsular Malaysia, which receives higher precipitation, while worse performances appear in the western and dryer Peninsular Malaysia.

There also have been some studies executed in the monsoon regions aiming to evaluate the applicability of satellite-based datasets in hydrologic simulations [53–60]. For example, Tong et al. [58] evaluated four satellite-based datasets (TRMM 3B42, TRMM 3B42RT, CMORPH RAW, and PERSIANN) through comparing with the gauged China Meteorological Administration dataset (CMA) in streamflow simulations over the Tibetan Plateau based on the distributed Variable Infiltration Capacity (VIC) hydrological model. It was found that the error sources of these datasets are systematically different in different seasons. Furthermore, TRMM 3B42 shows comparable performance to CMA for both monthly and daily streamflow simulations due to its monthly gauge adjustment. However, the other three satellite-based datasets only show potentials or little capability for streamflow simulations over TP. In addition, five satellite-based precipitation datasets (TRMM 3B42, TRMM 3B42RT, CMORPH RAW, CMORPH CRT, and CMORPH BLD) were used by Wang et al. [59] to simulate the daily streamflow by driving the distributed Vegetation Interface Processes (VIP) model over two river basins in the southeastern Tibetan Plateau. The results show that these satellite-based datasets perform better in summer than other seasons, and CMORPH BLD performs the best for runoff simulations. TRMM 3B42 and CMORPH CRT show much better performance than their uncorrected counterparts: TRMM 3B42RT and CMORPH RAW.

From the previous studies, we found that first, there are relatively few evaluations focusing on satellite-based precipitation datasets in the monsoon regions of southern China, which is a flood-prone area. Both the flood predictions and water resource management are mainly based on hydrological simulations. Moreover, most existing studies in the monsoon characterized regions only compare several commonly used satellite-based datasets (such as TRMM and CMORPH serial datasets) and some promising recently released precipitation datasets, such as PERSIANN CDR, Climate Hazards Group InfraRed Precipitation with Stations V2.0 (CHIRPS), and Multi-Source Weighted-Ensemble Precipitation V2.0 (MSWEP) have not been thoroughly evaluated. Second, it is crucial to ensure that the gauged benchmark reference is sufficient to reflect the real performance of satellite-based precipitation when testing satellite-based datasets. However, many studies compared the satellite-based datasets based on the sparse-gauge datasets or gridded datasets generating from sparse gauges, which may not accurately reflect the spatial characteristic of precipitation [47,49,54,59,60]. Furthermore, when evaluating the accuracies of satellite-based datasets, the gauged references in some studies are not independent of the satellite-based datasets, which uses the gauged precipitation as part of their source data [49,52,58].

Third, despite the fact that some studies show that the performance of hydrological simulation is highly dependent on the satellite-based datasets themselves in the monsoon regions, the uncertainties of hydrological models caused by different models' complexities could also influence the hydrological simulation. The impact of these two uncertainties has not been carefully examined.

The latest review article of Maggioni et al. [35] pointed out that one of the future research areas for satellite-based precipitation datasets is to study the conditions (climate type, basin area, acceptable error in the output, and model structure) under which satellite-based precipitation could be successfully used in hydrological models. In order to provide a comprehensive understanding of the error of the satellite-based precipitation and its error propagation through hydrological models for monsoon-characterized watersheds, this study tests the reliability of eight satellite-based precipitation datasets in hydrological modeling for a large-sized (>80,000 km^2) monsoon-characterized watershed (Xiangjiang River Basin) in southern China. Even though one of the main usages of the satellite-based datasets is for ungauged watersheds or watersheds with spare weather stations, the test of their reliability requires a watershed with dense gauges. The Xiangjiang River Basin, which has 267 precipitation gauges (referred to as the dense-gauge precipitation dataset in the study), can meet this requirement for an 80,000 km^2 surface area. All the eight satellite-based precipitation datasets include TRMM 3B42 (TRMM), PERSIANN, PERSIANN CDR, CMORPH RAW, CMORPH bias-corrected (CMORPH CRT), CMORPH gauge blended (CMORPH BLD), MSWEP, and CHIRPS. In addition to using the dense-gauge precipitation dataset as a reference, an independent gridded gauge-interpolated precipitation dataset is also used, which incorporates much fewer stations from the National Meteorological Information Center dataset from the China Meteorological Administration (CN05) [61]. As high-density gauged precipitation is usually not available in China, CN05 is commonly used for meteorological and hydrological studies over most watersheds [62–64]. This study could be extended to test whether CN05 is capable of being used as a reliable reference for using satellite-based datasets over other watersheds where gauges are much less dense. To investigate the uncertainty related to hydrological models, the lumped Xinanjiang (XAJ) model and the semi-distributed Soil and Water Assessment Tool (SWAT) model, with different complexities, are used.

2. Study Area and Datasets

2.1. Study Area

The Xiangjiang River Basin has a complex topography with elevation ranging from 0 to 2100 m above sea level and is located between 24.5°–28.1°N and 110.5°–114.0°E in the southern part of China (Figure 1). The Xiangjiang River originates from Haiyang Mountain in Guangxi province with a drainage area of 80,669 km^2 and a total length of 801 km, making it one of the largest tributaries of the Yangtze River [3,65]. The Xiangjiang River Basin, located in the subtropical and warm temperate zone, which is dominated by the East-Asian monsoon climate with heavy summer rainfall in the south, is an ideal experimental basin with a good relationship between precipitation and runoff [65]. The average temperature is around 17 °C, and the annual precipitation is close to 1500 mm with occasionally little snowfall in the winter. More than 70% of the annual precipitation occurs between March and August. In addition, there are abundant water resources in the Xiangjiang River Basin; the study of satellite-based precipitation could provide valuable information for flood forecasting and water resources management for the administrative department.

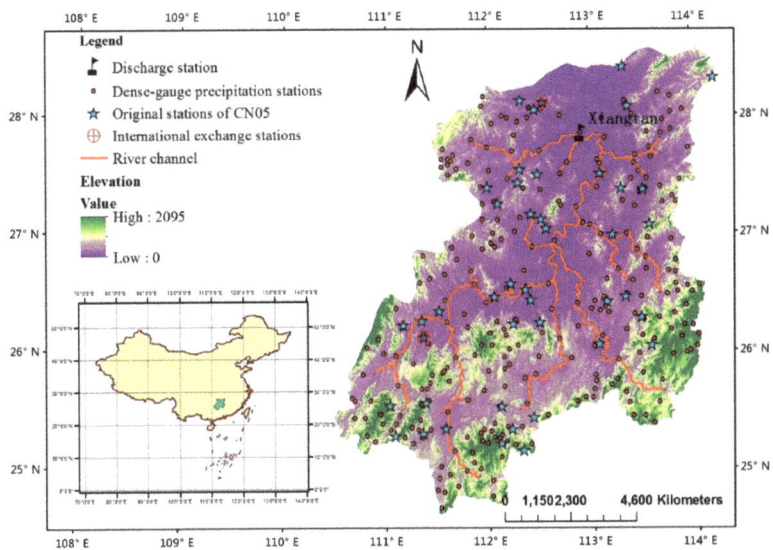

Figure 1. The location of Xiangjiang River Basin and its river channel, precipitation gauge stations (dense-gauge precipitation gauge stations, original stations of CN05, and international exchange stations) and discharge stations.

2.2. Data

In this study, eight satellite-based datasets are selected and can be further classified into three categories: (1) satellite-only (PERSIANN and CMORPH RAW), in which their quality fully depends on the raw satellite data, (2) satellite-gauge (TRMM, PERSIANN CDR, CMORPH CRT, and CMORPH BLD), in which their quality partly depends on gauge data, and (3) satellite-reanalysis-gauge/blended (MSWEP and CHIRPS), in which reanalysis data are blended. These datasets share the same spatial resolution of 0.25° × 0.25° for latitude and longitude, and the common period between 2003 to 2013.

Although PERSIANN and CMORPH RAW both incorporate PMW and IR to estimate rainfall, the proportion of PMW and IR is totally different between these two datasets. Specifically, CMORPH RAW is primarily based on PMW remote sensing of rainfall, while PERSIANN is mainly based on IR imagery [66,67]. Each satellite-gauge and blended (gauges, satellites, and reanalysis data) dataset blends different source data by using different data fusion methods. In general, CMORPH BLD and MSWEP directly incorporate daily gauge data, while TRMM and CMORPH CRT directly incorporate monthly gauge data. Unlike these four datasets specially designed to provide the best instantaneous accuracy, PERSIANN CDR (monthly precipitation) and CHIRPS (5-day precipitation) have been designed to achieve the best simulations of the most temporally homogeneous record.

Specifically, TRMM blended GPCC with their satellite-only counterparts TMPA 3B42RT (which, similar to CMORPH RAW, is also estimated primarily by PMW remote sensing of rainfall) by the inverse error variance weighting method [68]. CMORPH CRT was produced by blending the CMORPH RAW dataset with Climatic Prediction Center (CPC) and GPCC via the probability density function matching a bias correction method [69]. The optimal interpolation method was used to combine the CMORPH CRT with daily gauge analysis to produce the CMORPH BLD [69]. Instead of using gauged observations directly, PERSIANN CDR was adjusted to match the monthly satellite-gauge GPCP, which uses gauge-interpolated GPCC, to remove its monthly biases [6,70]. Although both MSWEP and CHIRPS are categorized as blended datasets, the data sources and fusion methods are totally different. MSWEP is mainly produced by giving weights to each dataset on each grid from different data sources (daily and monthly gauges such as CPC and GPCC, reanalysis from ERA-Interim,

Japanese 55-year Reanalysis (JRA 55) and satellite from CMORPH RAW, Global Satellite Mapping of Precipitation (GSMap MVK) and TRMM 3B42RT) based on their comparative performances at the surrounding gauges [71]. However, CHIRPS mainly uses the NOAA Climate Forecast System (CFS) reanalysis datasets to fill the missing values calculated by satellite datasets (from such as TRMM 3B42) and five-day gauged precipitation from datasets such as World Meteorological Organization's Global Telecommunication System [72]. More details of the above datasets are shown in Appendix A.

The reliability of the eight satellite-based precipitation datasets is evaluated by comparing it with two gauged precipitation datasets, including the dense-gauge dataset and the gridded gauge-interpolated dataset (CN05). As an important experimental basin, the Xiangjiang River Basin owns the dense-gauge precipitation dataset derived from a dense ground network of 267 precipitation stations with complete temporal coverage from 1963 to 2013, which is offered by the local hydrological department: the Water Conservation Bureau of Hunan Province. CN05, as a national gauge-interpolated dataset, is composed of daily precipitation estimates at the spatial resolution of 0.5° for the quasi-China coverage of 54°N to 18°S latitude from 1961 to 2016. CN05, independent from the dense-gauge precipitation dataset, is generated by blending daily precipitation data (2472 Chinese national weather gauges and 44 gauges locating in this study region) with Chinese mainland Digital Elevation Model (DEM) data (resampled from the Global 30 Arc Second Elevation Dataset, with a spatial resolution of 0.5° × 0.5°) using Thin Plate Spline algorithm (TPS) [73]. It is worth noting that CN05 is not independent of the eight satellite-based datasets. This is because two of the 44 gauges of CN05 in the study region are selected as the international exchange gauges that provide measured components (such as GPCC and CPC) from four satellite-gauges and two blended datasets. This means that the gauged components of the satellite-gauge and blended datasets come from the same source. In other words, factors that influence the performances of satellite-based datasets come from other data sources (satellite or reanalysis) or the blending strategies between and within various source data. Compared with the eight above-mentioned daily satellite-based precipitation datasets, which define a day as 0–23:59 UTC, both dense-gauge and CN05 precipitation datasets use the same daily precipitation time interval, from 8 UTC of one day to that of the next day. This ensures that the daily precipitation measurement in China, in the eastern eight zones, is executed simultaneously with daily precipitation measurements under the 0–23:59 UTC standard. A brief summary of the eight satellite-based datasets and two gauged datasets is presented in Table 1. The locations of 267 dense-gauge precipitation datasets, 44 precipitation gauges of source data of CN05, and two international exchange gauges are shown in Figure 1.

For hydrological modeling, temperature data from 13 stations and streamflow time series at the watershed outlet are also used. In addition, a Digital Elevation Model (DEM) dataset with a spatial resolution of 30 m, a land-use dataset with a spatial resolution of 1 km, and a soil dataset from Harmonized-world-soil-datasets (HWSD) are used to establish the semi-distributed SWAT model.

Table 1. Background information for selected precipitation datasets used in this study.

Dataset	Category	Spatial Resolution	Temporal Domain	Coverage	Developer	Link
Dense-gauge	Gauged	267 precipitation stations	1963–2013	-	the Water Conservation Bureau of Hunan Province	-
CMA	Gauge-interpolated	0.5° × 0.5°	1961–2016	54°N–18°S	China Meteorological Administration	http://data.cma.cn/data
TRMM	Satellite-gauge	0.25° × 0.25°	1998–present	50°N–50°S	NASA and Japan Aerospace Exploration (JAXA) Agency	ftp://trmmopen.gsfc.nasa.gov/pub/merged/3B42RT/
PERSIANN CDR			2003–present	60°N–60°S	the Center for Hydrometeorology and Remote Sensing (CHRS) at the University of California, Irvine	http://chrsdata.eng.uci.edu
CMORPH CRT			1998–present	50°N–50°S	Climate Prediction Center of NOAA	ftp://ftp.cpc.ncep.noaa.gov/precip/CMORPH_V1.0/CRT/0.25deg-DLY_00Z/
CMORPH BLD			1998–present	50°N–50°S	Climate Prediction Center of NOAA	ftp://ftp.cpc.ncep.noaa.gov/precip/CMORPH_V1.0/BLD/0.25deg-DLY_EOD/GLB/
CHIRPS	Blended		1981–present	50°N–50°S	the U.S. Geological Survey (USGS) Earth Resources Observation and Science (EROS) Center	ftp://ftp.chg.ucsb.edu/pub/org/chg/products/CHIRPS-2.0
MSWEP			1975–present	90°N–90°S	Hylke Beck in Princeton University	http://www.gloh2o.org
PERSIANN	Satellite-only		2003–present	60°N–60°S	the Center for Hydrometeorology and Remote Sensing (CHRS) at the University of California, Irvine	http://chrsdata.eng.uci.edu
CMORPH RAW			1998–present	50°N–50°S	Climate Prediction Center of NOAA	ftp://ftp.cpc.ncep.noaa.gov/precip/CMORPH_V1.0/RAW/0.25deg-DLY_00Z/

3. Methodology

The comparison of datasets is carried out in both precipitation evaluations and hydrological simulations. When evaluating the precipitation, we compared the differences among all satellite-based precipitation datasets on both areal mean and grid scales to better understand the hydrological impacts of the errors from the satellite-based datasets. This is because the areal mean precipitation and the spatial distribution of precipitation are respectively decisive factors in the lumped XAJ and semi-distributed SWAT models used in this study. When an evaluation is executed at the grid-scale, the dense-gauge observations are interpolated by the IDW method to 151 grids with a spatial resolution of 0.25° × 0.25°, which is the same with eight satellite-based precipitation data [74]. For CN05 with a spatial resolution of 0.5° × 0.5°, the precipitation in four 0.25° grids within one 0.5° grid shares the same value.

3.1. Hydrological Models

In this study, two hydrological models with different complexities, such as a conceptual lumped model and a physically-based semi-distributed model, are utilized for hydrological modeling. Both models have been successfully established in the Xiangjiang River Basin for many studies [3,44,75,76]. Compared to the lumped XAJ, which uses the areal mean precipitation as the model input, the semi-distributed SWAT uses precipitation from a single rain gauge closest to each sub-basin's centroid as the model input. Details of these two models are described below.

3.1.1. Xinanjiang Model (XAJ)

The XAJ model is a lumped conceptual rainfall–runoff model of a set of 15 variables developed in the 1970s [77,78]. It has been successfully used in humid regions of China [79–81]. Outflow simulation from the total outlet of the basin mainly consists of three phases: evapotranspiration, runoff generation, and runoff routing. Four parameters account for evapotranspiration, two account for runoff generation, and nine account for runoff routing. Its hydrological cycle is based on the water balance equation:

$$S_t + W_t = S_0 + W_0 + \sum_{i=1}^{t}(R_{day} - Q_{surf} - E_a - Q_{lat} - Q_{gw}), \quad (1)$$

where S_t and S_0 are the mean and initial free water storage capacity, W_t and W_0 are the mean and initial tension water storage, R_{day} is the amount of precipitation on day i, Q_{surf} is the amount of surface runoff on day i, E_a is the amount of evapotranspiration on day i, Q_{lat} is the amount of lateral flow on day i, and Q_{gw} is the amount of groundwater flow on day i.

The evapotranspiration is calculated by dividing the soil into three layers: an upper layer, a lower layer, and a deep layer. The storage curve calculates the total runoff according to the hypothesis that when the soil moisture content reaches the field capacity, all rainfall turns into a runoff. The rainfall exceeding infiltration is transformed into the surface runoff Q_{surf}, and the rainfall that has infiltrated belongs to the lateral flow Q_{lat} and groundwater flow Q_{gw}.

3.1.2. Soil and Water Assessment Tool Model (SWAT)

SWAT, a physically-based semi-distributed model, is designed to predict the effects of land management practices on the hydrology, sediment, and contaminant transport [82]. SWAT could be operated under different soil compositions, land uses, and management conditions in an agricultural watershed [3,83]. Different from the XAJ model, which uses the whole basin as the operation unit, SWAT divides the entire basin into several unit basins, and each unit basin is further divided into several Hydrologic Research Units (HRUs). Each HRU is calculated individually based on relatively homogeneous land use, land cover, and soil types. The water balance of SWAT is described below as:

$$SW_t = SW_0 + \sum_{i=1}^{t}\left(R_{day} - Q_{surf} - E_a - W_{seep} - Q_{gw}\right), \quad (2)$$

where SW_t is the final soil water content, SW_0 is the initial soil water content on day i, t is the time, R_{day} is the precipitation amount on day i, Q_{surf} is the surface runoff amount on day i, and W_{seep} is the water amount entering the vadose zone from the soil profile on day i.

The Penman–Monteith method is used to estimate evapotranspiration E_a [84]. The surface runoff volume Q_{surf} is calculated by a Soil Conservation Service Curve Number method, and groundwater flow Q_{gw} is simulated by creating a shallow aquifer. The outlet simulation of basin is calculated by the Muskingum method for each sub-basin's simulation results [85].

3.1.3. Model Calibration and Validation

XAJ and SWAT models are respectively calibrated using the Shuffled Complex Evolution (SCE-UA) algorithm [86] and Sequential Uncertainty Fitting version 2 (SUFI2) algorithm [87], using the Nash–Sutcliffe efficiency (NSE shown in Table 2) coefficient as the objective function. Two models are calibrated from 2004 to 2010 and validated from 2011 to 2013, and 2003 is used as the spin-up year.

3.2. Statistical Analysis Methods

A set of statistical indices is utilized to evaluate the performance of eight satellite-based datasets in preserving precipitation and simulating watershed runoff. For precipitation evaluation, the indices include (1) four categorical statistics for detection error, (2) three quantitative metrics, of which two of them could reflect the systematic and random errors, and (3) four extreme precipitation statistics. There is one metric for hydrological evaluation to determine the overall hydrological performances and three hydrological statistics to reflect the characteristic values for streamflow. Additionally, the error propagation from precipitation to streamflow is qualified by two absolute ratios. A list of the indices can be found in Table 2, and more details are explained in the following section.

3.2.1. Precipitation Indices

Detection, systematic, and random errors are three main error sources of satellite-based datasets [35,88]. False alarms (when gauges do not observe the satellite-detected precipitation) and missed rain (when the gauge-observed precipitation are not actually detected by satellites) constitute the detection errors [89]. When the satellite correctly detects precipitation, errors of estimated precipitation compose systematic and random errors [90–93].

In this study, four categorical statistics: the frequency bias index (FBI), the probability of detection (POD), the false alarm ratio (FAR), and the equitable threat score (ETS) are used to quantify the detection errors of each satellite-based dataset [1]. The FBI reflects the tendency to underestimate or overestimate rainfall events. The FAR (POD) measures the fraction of false alarms (rain occurrences) that were correctly detected. The ETS provides an overall skill measurement of the correctly detected rain events (observed and/or detected).

The three quantitative statistics of precipitation are the relative bias (RB), unbiased root mean squared error (ubRMSE), and the coefficient of determination (R^2). RB reflects the systematic error, which is the relative difference in the long-term mean values of the two series. Although RMSE shows the amplitude of differences between the two series, it could not directly reflect the random error unless the system error is removed by subtracting the mean difference from the RMSE to get the ubRMSE. R^2 indicates the correlation between two series.

Four extreme statistics are selected from the recommended list by the joint World Meteorological Organization Commission for Climatology/World Climate Research Programme project on Climate

Change Detection and Indices (https://www.climdex.org/indices.html). These are the annual total precipitation when daily precipitation amount on a wet day > 99th percentile (R99pTOT), the annual daily precipitation amount on a wet day (SDII), the maximum length of wet and dry spells (CWD and CDD). P99pTOT is one threshold index, and SDII reflects the intensity of extreme precipitation. CWD (CDD) shows the duration of extreme precipitation (non-precipitation) events.

3.2.2. Hydrological Indices

The widely used metrics NSE is used to evaluate the performance of each precipitation dataset for hydrological simulations. NSE is calculated as the ratio of residual variance to measured discharge variances [94]. Simulated discharges using these datasets were also compared against their gauged counterparts using three hydrological statistics: daily mean discharge, winter low flow (5th percentile of the winter flow), and summer high flow (95th percentile of the summer flow).

3.2.3. Error Propagation Indices

Two absolute ratios (γ) between error metrics (RB and ubRMSE) for the runoff and precipitation series are used to quantify the error propagation through the precipitation–runoff process. γ_{RB} and γ_{ubRMSE} respectively reflect the systematic and random error propagation effects. They are always greater than 0 due to their absolute values, and values larger (smaller) than 1 indicate the amplification (dampening) of the error from precipitation to runoff.

Table 2. List of statistical indexes used in this study.

Category		Index	Equation/Description	Range and Optimal Value
Precipitation indices	Categorical statistics	FBI	$\frac{a+b}{a+c}$	$(0, \infty)$, 1
		POD	$\frac{a}{a+c}$	$(0, 1)$, 1
		FAR	$\frac{b}{a+b}$	$(0, 1)$, 0
		ETS	$\frac{a+H_e}{a+b+c-H_e}$ ($H_e = \frac{(a+b)(a+c)}{N}$ where N is the total number of estimates)	$(-\infty, 1)$, 1
	Quantitative metrics	RB	$\frac{\sum_{i=1}^{i=n}(S_i-G_i)}{\sum_{i=1}^{n} G_i}$	$(-\infty, \infty)$, 0
		ubRMSE	$\sqrt{\frac{\sum_{i=1}^{i=n}(G_i-S_i)^2}{n} - \left(\frac{\sum_{i=1}^{n}(G_i-S_i)}{n}\right)^2}$	$(0, \infty)$, 0
		R^2	$\frac{\left[\sum_{i=1}^{i=n}(G_i-\bar{G})(S_i-\bar{S})\right]^2}{\sum_{i=1}^{i=n}(G_i-\bar{G})^2 \sum_{i=1}^{i=n}(S_i-\bar{S})^2}$	$(0, 1)$, 1
	Extreme statistics	R99pTOT	Annual total precipitation when daily precipitation amount on a wet day>99th percentile	-
		SDII	Annual daily precipitation amount on wet day	-
		CWD	Maximum length of wet spell, maximum number of consecutive days with daily precipitation ≥ 1 mm	-
		CDD	Maximum length of dry spell, maximum number of consecutive days with daily precipitation < 1 mm	-
Hydrological indices	Evaluation metrics	NSE	$1 - \left[\frac{\sum_{i=1}^{n}(Y_i^{obs}-Y_i^{sim})^2}{\sum_{i=1}^{n}(Y_i^{obs}-\bar{Y})^2}\right]$	$(-\infty, 1)$, 1
	Hydrological statistics	DMD	Daily mean discharge	-
		WLF	Winter low flow (5th percentile)	-
		SHF	Summer high flow (95th percentile)	-
Error propagation indices		γ_{RB}	$\frac{\text{RB of runoff}}{\text{RB of precipitation}}$	$(-\infty, \infty)$, -
		γ_{ubRMSE}	$\frac{\text{ubRMSE of runoff}}{\text{ubRMSE of precipitation}}$	$(0, \infty)$, -

4. Results and Discussion

4.1. Precipitation Evaluation

4.1.1. Seasonal Patterns of Precipitation Datasets

Figure 2 presents the seasonality (spring: March–May, summer: June–August, autumn: September–November, winter: December–February; wet season: April–September and dry season: October–March) of the mean precipitation for all ten precipitation datasets (eight satellite-based precipitation datasets, one gauged precipitation (i.e., the dense-gauge dataset), and one gauge-interpolated precipitation (i.e., CN05)). All stations or grids within the watershed are averaged to a single time series to calculate the seasonal mean values. The figure graphically demonstrates that CN05 agrees well with the dense-gauge observation for all four seasons. Specifically, CN05 presents a small RB within ±7.0% for seasonal precipitation (−2.4% for spring, −6.1% for summer, −1.7% for autumn, and 0.3% for winter). With the exception of satellite-only datasets, which considerably underestimate the precipitation for all seasons, the satellite-based datasets also reasonably represent the observed seasonality. However, all of them are worse than CN05 for all seasons. The better performance of PERSIANN CDR among satellite-based datasets for seasonal precipitation, especially in spring, summer, and autumn, could reflect the effects of its blending strategies. PERSIANN CDR maintains monthly precipitation that is consistent with the monthly GPCP, and GPCP is mainly composed of gauged precipitation datasets (e.g., GPCC) [70]. In addition, all the satellite-gauge datasets overestimate the dense-gauge precipitation in summer and the wet season while underestimating in winter. In addition, both blended datasets (MSWEP and CHIRPS) overestimate the precipitation all year round. TRMM, CMORPH BLD, and MSWEP fit the dense-gauge precipitation better in the dry season than the wet season, while PERSIANN CDR, CMORPH CRT, CHIRPS, and satellite-only datasets perform better in the wet season than the dry season.

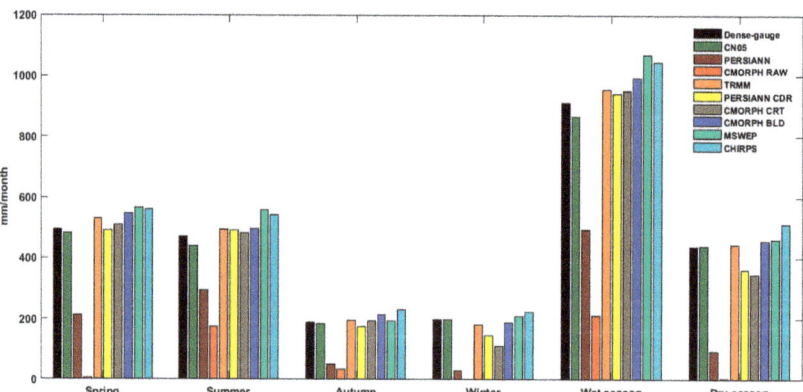

Figure 2. Mean seasonal precipitation (spring: March–May, summer: June–August, autumn: September–November, winter: December–February, dry season: October–March and wet season: April–September) of 2003–2013 from the dense-gauge precipitation and eight gridded datasets (National Meteorological Information Center dataset from the China Meteorological Administration (CN05), Precipitation Estimation from Remotely Sensed Information Using Artificial Neural Networks (PERSIANN), Climate Prediction Center MORPHing-raw satellite dataset (CMORPH RAW), Tropical Rainfall Measuring Mission (TRMM), PERSIANN Climate Data Record (PERSIANN CDR), CMORPH bias-corrected (CMORPH CRT), CMORPH gauge blended (CMORPH BLD), Multi-Source Weighted-Ensemble Precipitation V2.0 (MSWEP) and Climate Hazards Group InfraRed Precipitation with Stations V2.0 (CHIRPS)).

The spatial distributions of summer precipitation are also presented for all datasets in Figure 3. The dense-gauge datasets are presented as color dots, while all the other datasets are presented as grids. Generally, summer precipitation is heavier in high elevation areas (southeastern, southwestern, and southern parts) than in other regions. CN05 clearly missed quite some regional intensive precipitation (such as the heavy precipitation in the southeastern parts of the region), which can even be captured by MSWEP and CHIRPS. The bad performance of CN05 may be caused by two reasons: (1) its lower spatial resolution (0.5° × 0.5°) and (2) its less gauged source data compared with the dense-gauged dataset. Satellite-only datasets underestimate precipitation for all grids, even though PERSIANN can capture the heavy precipitation signal in mountain areas. Although all satellite-gauge datasets could capture this spatial distribution pattern, these datasets still underestimate the heavy precipitation in mountain areas (southern and southeastern parts) while overestimating the small precipitation in central plain regions. The spatial distributions of winter precipitation, as shown in Appendix B in Figure A1, display similar patterns, as CN05 still performs relatively worse than the two blended datasets: MSWEP and CHIRPS. The better performance of blended datasets for the spatial distribution of seasonal precipitation may be due to their reanalysis components.

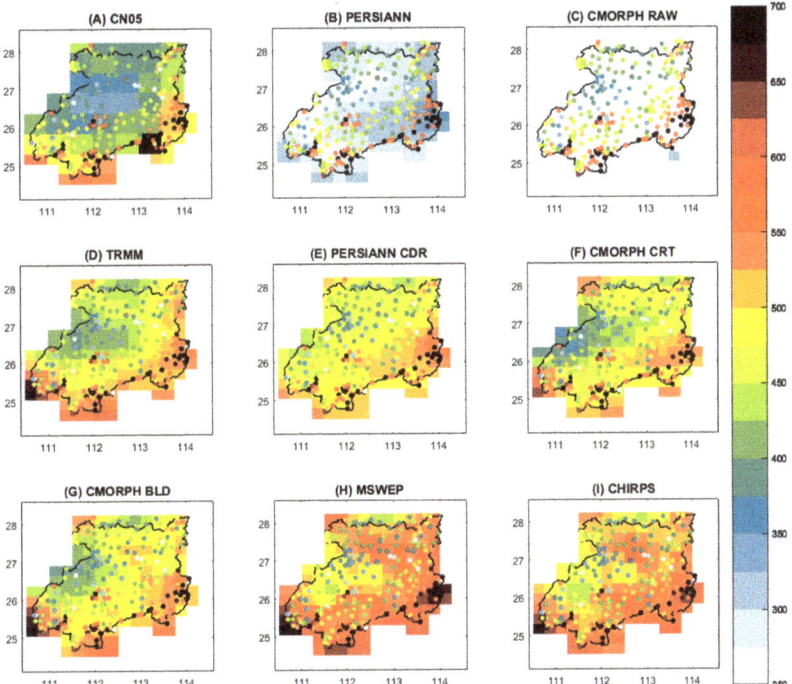

Figure 3. Summer precipitation (mm) during June–August of 2003–2013 from CN05 and eight satellite-based datasets compared with the dense-gauge dataset, which is shown as colored dots.

4.1.2. Error Structures of Precipitation Datasets

Figure 4A presents two types of the daily gridded precipitation information for the dense-gauge, CN05, and eight satellite-based datasets: (1) bar charts represent the frequency distribution of precipitation under seven different rain rate classes (0, 0–1, 1–5, 5–10, 10–25, 25–50 and >50 mm/day) and (2) line charts represent the contribution of the precipitation amount under each rain rate class to the total precipitation. As shown in the bar charts, PERSIANN CDR, CMORPH CRT, and CMORPH BLD are close to the gauged counterparts of where the precipitation frequencies decrease from 0 to

0–1 mm and slightly increase under the 1–5 mm class, and then decrease until the >50 mm class. These tendencies of precipitation frequency under the 0, 0–1, and 1–5 mm classes are inaccurately represented by MSWEP. Another two satellite-based datasets (TRMM and CHIRPS) overestimate the frequencies of no rain (0 mm) and heavy rain (>50 mm) and underestimate little rain (0–1 mm). Line charts show that the largest precipitation contribution of all datasets except for TRMM and CMORPH CRT occurs at the 10–25 mm class. Large differences in the precipitation contribution among datasets occur at the 25–50 mm and >50 mm classes.

The detection errors of each satellite-based dataset are quantified based on the FBI, FAR, POD, and ETS in terms of the 11-year (2003–2013) annual, wet season, and dry season precipitation processes. Figure 4B presents the distribution of the FBI for nine precipitation datasets (CN05 and eight satellite-based datasets). FBI values of CMORPH RAW at the 25–50 (13.66) and >50 (89.35) intervals being larger than 6 are not demonstrated, which is the same as Figure 4C,D. Although both satellite-gauge and blended datasets poorly simulate the annual FBI values in the rain rates of 0 mm (e.g., the FBI of MSWEP is 2.55) and 0–1 mm (e.g., the FBI of TRMM is 3.95), they overall outperform the satellite-only categories, which have worse annual FBI results under more than half of the rain rate classes (Figure 4B). Figure 4C,D further demonstrates that more overestimations of FBI values of satellite-only datasets under most rain rate classes (5–10, 10–25, 25–50, and >50 mm) occur in the dry season than the wet season. The larger underestimation of precipitation events in the dry season is in good agreement with the seasonal precipitation amount in Section 4.1.1 and could further explain the sources of poor performances for satellite-only datasets. This may be because of the underestimation of precipitation events with the rain rate classes being larger than 10 mm during the wet season and the underestimation of all precipitation events during the dry season. As the rain rate class increases, the FBI of satellite-gauge datasets improves until the precipitation class exceeds 50 mm for both seasons. The annual FBI values of CMORPH CRT (0.78), TRMM (0.62), and CHIRPS (0.49) at this class are less than 1, indicating that these datasets overestimate the number of heavy rain events. This may also explain the overestimation of the percentage of heavy rains (Figure 4A).

FAR, POD, and ETS of satellite-based datasets also show obvious seasonal patterns. Two satellite-only datasets significantly deteriorate with the increasing rain rate classes in terms of the annual FBI, FAR, and POD, indicating their inability to capture the heavy precipitation. These two datasets clearly perform better in the wet season than in the dry season, especially in terms of the POD (Figure 4I,J) and ETS (Figure 4L,M). However, CMORPH BLD and MSWEP show an opposite seasonal pattern, as the better performance occurs in the dry season than the wet season in terms of the three statistics. In addition, both of them maintain their superiority among all the satellite-gauge datasets. Although gauge-interpolated CN05 shows relatively worse performances than CMORPH BLD and MSWEP, it shares similar seasonal patterns with them and also outperforms the other six satellite-based datasets with regard to POD and ETS under the most rain rate classes (0–1, 1–5, 5–10, and 10–25 mm).

Figure 5 shows the RB, ubRMSE, and R^2 of nine precipitation datasets (CN05 and eight satellite-based datasets) at both grid (shown as boxplots) and watershed-average scales (shown as radar plots). Generally, the performances of each precipitation dataset under two different scales are basically consistent in terms of all three quantitative statistics. Figure 5A,B show that CN05 presents a better RB than the eight satellite-based datasets. Among all satellite-gauge datasets, TRMM, PERSIANN CDR, and CMORPH CRT show the smallest RBs, indicating their smaller systematic errors, at both grid and watershed-averaged scales. CMORPH BLD and two blended datasets (MSWEP and CHIRPS) generally show positive RB under both scales, especially for CHIRPS, which overestimates the daily precipitation for more than 86.1% of the grids and has an RB of 16.5% at the watershed-average scale. In contrast, satellite-only datasets considerably underestimate the mean precipitation at both scales.

Random errors of CN05 and satellite-based datasets are quantified using the ubRMSE (Figure 5C). CN05 shows relatively larger random errors than the satellite-based datasets except for CMORPH RAW and CHIRPS. In addition, large differences are observed among satellite-based datasets. Specifically, CMORPH BLD presents the smallest ubRMSE with the median value of 6.31 mm at the grid scale

(Figure 5C) and 2.09 mm at the watershed-average scale (Figure 5D), while CHIRPS presents the largest ubRMSE with the median value of 10.94 mm at the grid scale and 6.28 mm at the watershed-average scale. MSWEP performs the best among all nine precipitation datasets with a median value of 5.83 mm at the grid scale and 2.14 mm at the watershed-average scale.

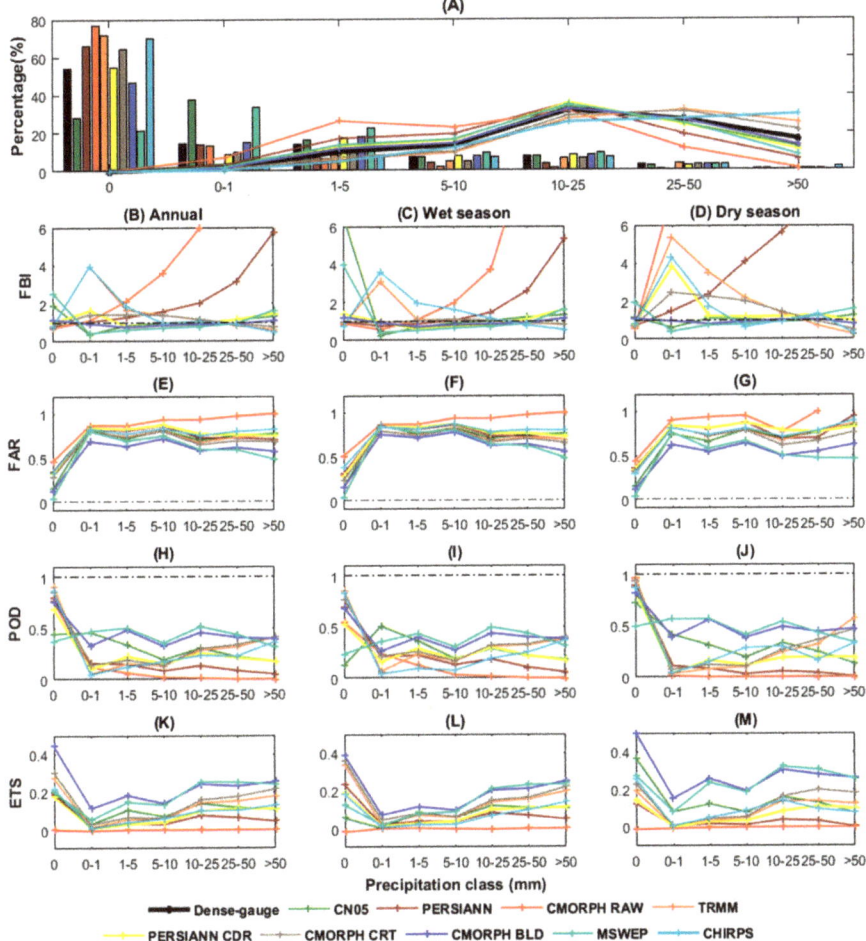

Figure 4. Frequency distribution of daily rainfall (**A**, shown as bar graph), contribution of rain rate classes to the annual accumulation (**A**, shown as line chart) as well as frequency bias index (FBI), false alarm ratio (FAR), probability of detection (POD), and equitable threat score (ETS) values of the 11-year annual (**B,E,H,K**), wet season (**C,F,I,L**), and dry season (**D,G,J,M**) precipitation process of 2003–2013 on a grid scale for seven daily precipitation thresholds over the Xiangjiang River Basin for the nine gridded datasets (CN05, TRMM, CHIRPS, PERSIANN CDR, CMORPH CRT, CMORPH BLD, MSWEP, PERSIANN and CMORPH RAW) and the dense-gauge dataset.

Figure 5E,F presents the R^2 values for all nine precipitation datasets, and both clearly reflect the influence of the blending methods and the incorporated gauged datasets on R^2. Datasets designed to provide the best instantaneous accuracy of precipitation (TRMM, CMORPH CRT, CMORPH BLD, and MSWEP) perform relatively better than those aimed to achieve the most temporally homogeneous record (PERSIANN CDR and CHIRPS). Within the four better-behaved satellite-based datasets, those that directly incorporate daily gauge data (CMORPH BLD and MSWEP) clearly perform better than those that directly incorporated monthly gauge data (TRMM and CMORPH CRT). Two satellite-only datasets show the worst performance among all the satellite-based datasets. Similarly, CN05 is also less correlated with the dense-gauge dataset than half of the satellite-based datasets (TRMM, CMORPH CRT, CMORPH BLD, and MSWEP).

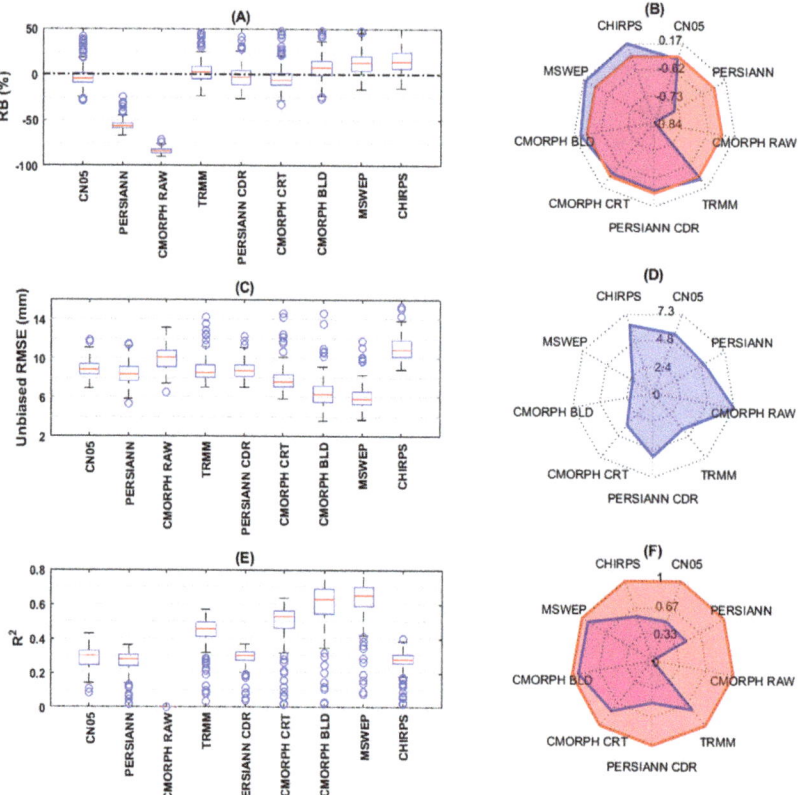

Figure 5. Relative bias (RB), unbiased root mean squared error (RMSE), and R^2 of the daily precipitation for the nine gridded datasets on both grid (shown as boxplot in (**A**,**C**,**E**)) and watershed (shown as radar plot in (**B**,**D**,**F**)) scales. In radar plots at right sides, red and blue lines represent the optimal values (RB (0), unbiased RMSE (0) and R^2 (1)) and the results of gridded datasets of each statistic, respectively.

4.1.3. Simulation of Extreme Precipitation

The results of four extreme precipitation statistics are presented in Figure 6 for nine datasets (CN05 and eight satellite-based datasets) at both grid (shown as relative bias compared to the dense-gauge precipitation dataset in boxplots) and watershed (shown as the absolute value in radar plots) scales. In the radar plots, red and blue lines represent the results of the dense-gauge and each dataset, respectively.

R99pTOT (Figure 6A,B) reflects the total precipitation of heavy rain. The R99pTOT values of satellite-gauge and blended datasets except PERSIANN CDR and CHIRPS are similar to the dense-gauge observation, especially for more than 50% grids having biases within ±20.0% at the grid scale. Specifically, PERSIANN CDR underestimates R99pTOT with more than 56.3% of grids having negative bias being smaller than −20% and a relative bias of 9.3% at the watershed scale.

However, CHIRPS overestimates R99pTOT at both scales (with more than 63.6% of grids having positive bias being larger than 20%, and a relative bias of 45.3% at the watershed scale). Additionally, two satellite-only datasets underestimate R99TOT.

The SDII values are shown in Figure 6C,D, and the similar results of SDII and R99pTOT can be explained by two factors: (1) heavy precipitation accounts for a large proportion of the annual precipitation amount, and (2) the number of wet days is similar for all nine datasets. The CWD is presented in Figure 6E,F at the grid and watershed-average scales, respectively. The CDD is presented in Figure 6G,H. The results show that CMORPH BLD maintains its superiority among all the datasets in simulating these two extreme statistics. However, the other seven satellite-based datasets could not accurately capture the CDD and the CWD at the same time, especially for the CDD, which is used as a criterion for representing droughts. For example, CMORPH CRT shows a small bias of CWD at both grid (with more than half of the grids have a bias of between ±10.0%) and watershed-average scales (CMORPH CRT: 18 days and the dense-gauge: 19 days). On the contrary, the CDD of CMORPH CRT is not accurately estimated with more than 50.0% of the grids having a bias larger than 10.0% and bias of 44.7% at the watershed-average scale (CMORPH CRT: 55 days and the dense-gauge: 38 days).

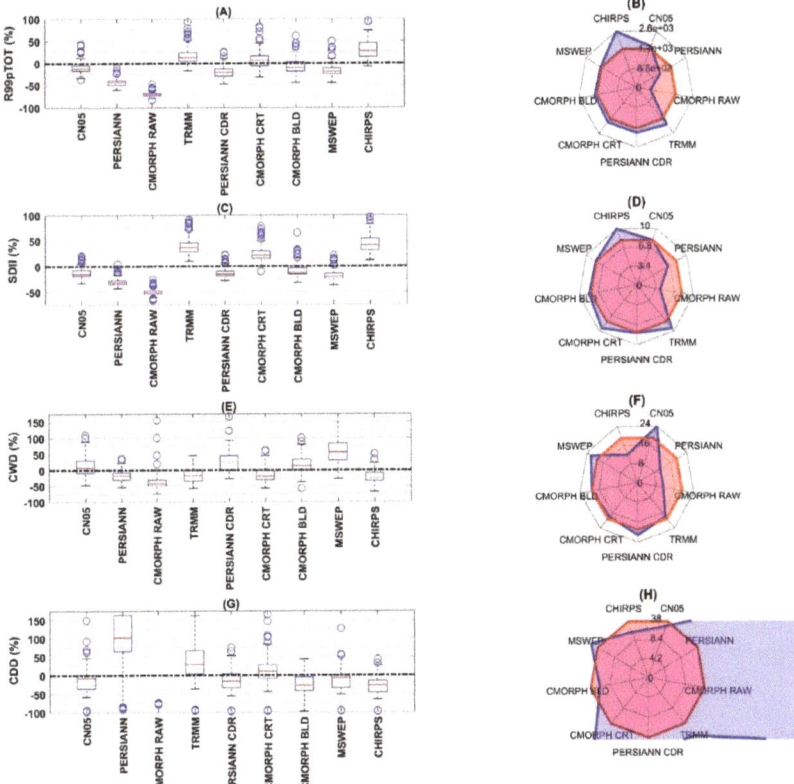

Figure 6. Annual total precipitation when daily precipitation amount on a wet day > 99th percentile (R99pTOT), the annual daily precipitation amount on a wet day (SDII), and the maximum length of wet and dry spells (CWD and CDD) of the daily precipitation for the nine gridded datasets on both grids (shown as boxplot in (**A,C,E,G**)) and areal mean (shown as radar plot in (**B,D,F,H**)) scales. In radar plots at right sides, red line and blue lines respectively represent the results of the dense-gauge dataset and the nine other gridded datasets.

CN05 better represents these four extreme precipitation statistics than all satellite-based datasets, especially for CDD, as shown in Figure 6G,H. It also shows tiny detection and systematic errors in the previous comparison in Section 4.1.2. However, CN05 misses quite some regional intensive seasonal precipitation and has larger random errors and worse R^2 compared with more than half of the satellite-based datasets. In other words, the bad performance of CN05 indicates that some satellite-based datasets are effective in representing the spatial distribution of precipitation. However, this effect can be missed by gauge-interpolated datasets using sparse gauges with a relatively coarser spatial resolution. Therefore, there is a risk of having CN05 as the reference when investigating the statistical properties of satellite-based precipitation, especially for high-precision datasets.

4.2. Hydrological Simulations

Eight satellite-based datasets and gauge-interpolated CN05 are further compared against the dense-gauge dataset in hydrological modeling by both XAJ and SWAT models calibrated by observed streamflow. Both models are adequately calibrated with NSE values of 0.89 (XAJ) and 0.86 (SWAT) for calibration, and 0.89 (XAJ) and 0.84 (SWAT) for validation (Table 3).

Table 3. Comparison of Nash–Sutcliffe efficiency (NSE) of both Xinanjiang (XAJ) and Soil and Water Assessment Tool (SWAT) models in daily step simulation based on the dense-gauge and the nine precipitation datasets.

Datasets	Period	XAJ NSE	SWAT NSE
Dense-gauge	Calibration (2004–2010)	0.89	0.86
	Validation (2011–2013)	0.89	0.84
CN05	(2004–2013)	0.86	0.83
PERSIANN		−0.4	−0.31
CMORPH RAW		−0.97	−0.95
TRMM		0.73	0.72
PERSIANN CDR		0.56	0.58
CMORPH CRT		0.75	0.73
CMORPH BLD		0.84	0.81
MSWEP		0.78	0.79
CHIRPS		0.44	0.48

For illustrating the intra-annual variability of the hydrological process, Figure 7 shows the mean monthly hydrographs of observed and the simulated streamflow of the dense-gauge and the other nine precipitation datasets based on two models. The reason for using a monthly hydrograph rather than a daily hydrograph is to avoid noises when calculating the climatology due to the relatively short time period (i.e., 10 years) [95,96]. It can be observed that (1) the most precise simulation of discharge is achieved by the gauge-interpolated CN05 among all nine precipitation datasets. CMORPH BLD, MSWEP, TRMM, and CMORPH CRT offer better performance than the other satellite-based datasets. CHIRPS and PERSIANN CDR, respectively, overestimate and underestimate the observed discharge for almost the whole year. (2) During the flood periods (from April to August), the simulation processes of both the dense-gauge and the other nine datasets based on the XAJ model are obviously larger than results based on the SWAT model. Similar results were also discovered by Xu et al. [3], who used XAJ and SWAT models to test the ability of two reanalysis datasets in simulating flood events in the Xiangjiang River Basin.

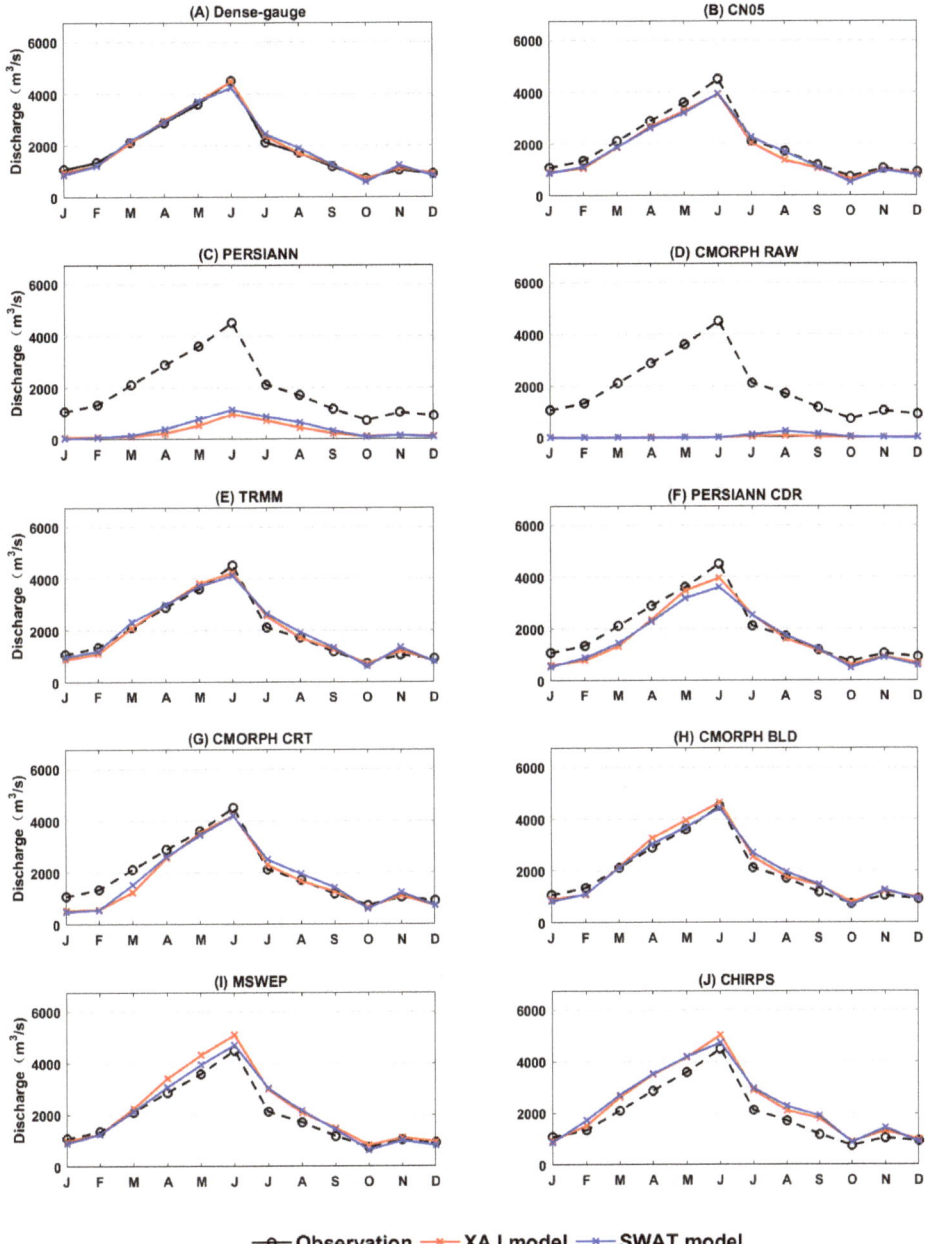

Figure 7. Simulation results of mean monthly hydrographs (2004–2013) using nine gridded datasets and the dense-gauge dataset in the SWAT together with XAJ models.

To further quantify the performance of satellite-based datasets in representing streamflow time series, the NSE values of two hydrological models based on daily streamflow are calculated and presented in Table 3. Results based on the XAJ model show that three satellite-gauge precipitation datasets (TRMM, CMORPH CRT, and CMORPH BLD) and blended datasets (MSWEP) are satisfactory in simulating streamflow time series, with NSE being larger than 0.72. CMORPH BLD (NSE = 0.84) outperforms all other satellite-based datasets. CHIRPS (NSE = 0.44) and PERSIANN CDR (NSE = 0.56) show moderate performances. Satellite-only datasets cannot represent the observed streamflow time series with NSE = −0.97 for CMORPH RAW and NSE = −0.40 for PERSIANN. The semi-distributed SWAT shows the similar daily simulation performance of each dataset with the lumped XAJ, but the performance of satellite-gauge datasets in SWAT is slightly worse than that in the XAJ except for PERSIANN CDR. Similar to PERSIANN CDR, both blended datasets, CHIRPS (NSE = 0.44/0.48 for XAJ/SWAT) and MSWEP (NSE = 0.78/0.79 for XAJ/SWAT), perform better in SWAT than in XAJ. Despite some differences in the simulation performances of the two models, the relative sort orders of the datasets based on NSE are almost consistent in both models. The best simulation was achieved by satellite-gauge CMORPH BLD, which was followed by blended MSWEP, TRMM, and CMORPH CRT. CHIRPS. PERSIANN CDR performed moderately; however, satellite-only datasets showed the worst performance. This consistency indicates that using different models does not significantly alter the relative performances of streamflow simulation of satellite-based precipitation datasets.

Three hydrological statistics (daily mean discharge, winter low flow, and summer high flow) are further used to compare the daily simulated discharge of both the dense-gauge and the nine alternative datasets against their observed counterparts. Figure 8 presents the annualized results (shown as the relative bias between the simulated discharge of each precipitation dataset and the observed discharge) of three statistics from 2004 to 2013. XAJ and SWAT models show similar results for daily mean discharge (Figure 8A); however, SWAT obviously underestimates the other two hydrological statistics (Figure 8B,C), especially for the winter low flow. Based on three statistics, CMORPH BLD consistently performs better than other satellite-gauge datasets. PERSIANN CDR and CHIRPS respectively underestimate and overestimate the observed discharge for both models. Blended MSWEP performs well, although its daily maxima discharge in the XAJ model shows an obvious overestimation (the results of 8 years are more than 0) and underestimation in the SWAT model (the results of 7 years are less than 0). Similar to previously used indexes, satellite-only datasets still show the worst performance.

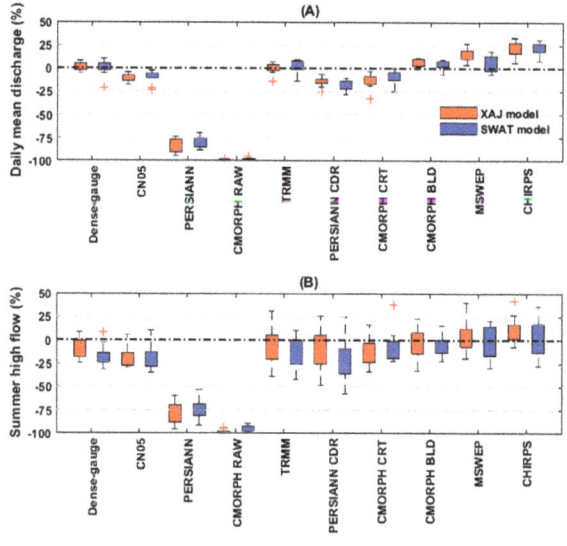

Figure 8. *Cont.*

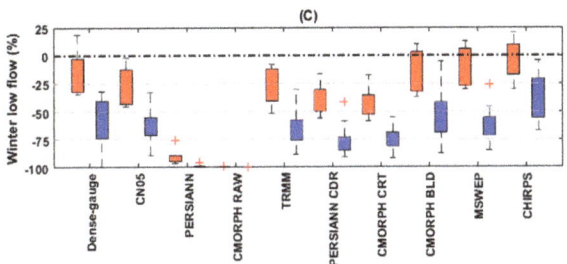

Figure 8. Boxplot of the relative bias of the daily mean discharge (**A**), summer high flow (**B**), and winter low flow (**C**) simulated using the nine gridded datasets and the dense-gauge dataset based on SWAT (red) together with XAJ (blue) models. Each boxplot is constructed with one value from each year of 2004–2013.

4.3. Error Propagation

RB and ubRMSE of streamflow respectively reflect the systematic and random errors of each dataset in simulating the streamflow. Figures 9 and 10 respectively show the RB and ubRMSE of annual, wet, and dry seasons streamflow and their corresponding propagation factors (γ_{RB} and γ_{ubRMSE}) simulated using CN05 and eight satellite-based precipitation datasets from 2004 to 2013.

4.3.1. Systematic Error Propagation

Generally, TRMM, with the minimal RB (systematic error) of annual streamflow performs the best among all datasets, which is then followed by CMORPH CRT and CMORPH BLD, displaying comparable performance with CN05 (Figure 9A). Two satellite-only datasets considerably underestimate the annual streamflow. However, their results of the systematic error propagation factor (γ_{RB} shown in Figure 9B) are larger than 1, indicating amplification of the systematic error when translating the precipitation into a runoff. TRMM, PERSIANN CDR, and CHIRPS have the same amplified effect for the systematic error of the precipitation, while γ_{RB} values for the other five datasets are around 1.

There is a seasonal trend for the RB of streamflow for all datasets in which the range of RB values for the wet season streamflow (Figure 9C) is much smaller than that for the dry season (Figure 9E). This narrow RB range means a smaller inter-annual difference in the wet season. As for RB, six out of nine datasets (all datasets except CMORPH BLD, MSWEP, and CHIRPS) show smaller RBs (closer to 0) in the wet season than in the dry season. Thus, the more apparent amplification of the systematic error of precipitation to runoff (the larger results of γ_{RB}) occurs in the dry season compared to the wet season for nearly all nine datasets except for satellite-only datasets (Figure 9D,F).

Moreover, the hydrological models also influence the RB of streamflow. During the wet season, SWAT generally performs much better than XAJ for more than half of the datasets (all datasets except for TRMM, PERSIANN, and CHIRPS, Figure 9C). While during the dry season, XAJ outperforms SWAT for more than half of the datasets (all datasets except for CMORPH CRT and MSWEP, Figure 9E).

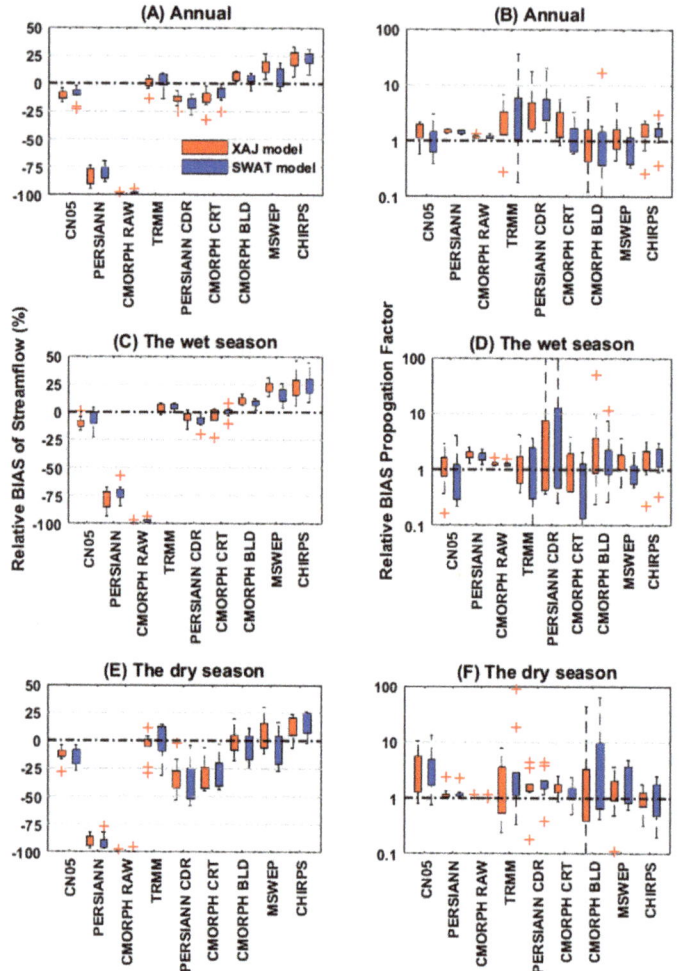

Figure 9. Relative bias of streamflow (**A,C,E**) and relative bias propagation factor (**B,D,F**) of the whole year, the wet season and the dry season for the nine gridded datasets based on SWAT (red) together with XAJ (blue) models. Each boxplot is constructed with one value of each year from 2004 to 2013.

4.3.2. Random Error Propagation

Figure 10A demonstrates that the streamflow values of ubRMSE are not distinctive among nine datasets, except for two satellite-only data, which have significantly larger values. Among the rest of the datasets, PERSIANN CDR and CHIRPS show the largest random errors of streamflow. CN05 shows the minimum ubRMSE of streamflow; however, this is different from its relatively larger ubRMSE of the precipitation (as demonstrated in Section 4.1.2). This discrepancy in the ubRMSE of precipitation and streamflow for CN05 is due to its largest dampening effect of random error. All the other eight datasets have similar dampening effects with γ_{ubRMSE} being smaller than 1 (Figure 10B), and CMORPH BLD along with MSWEP have the largest γ_{ubRMSE} among these datasets.

ubRMSE of streamflow also has a seasonal trend with its values and ranges in the wet season (Figure 10C) being larger than those in the dry season (Figure 10E) for all datasets. This seasonal difference also applies to the random error propagation factor γ_{ubRMSE} (Figure 10D,F). ubRMSE of the same precipitation dataset generated from different hydrological models is different (Figure 10C,E).

Specifically, SWAT generates a larger ubRMSE than that of the XAJ model for nearly all datasets (nine datasets except for PERSIANN CDR and CHIRPS), especially during the wet season (Figure 10D,F).

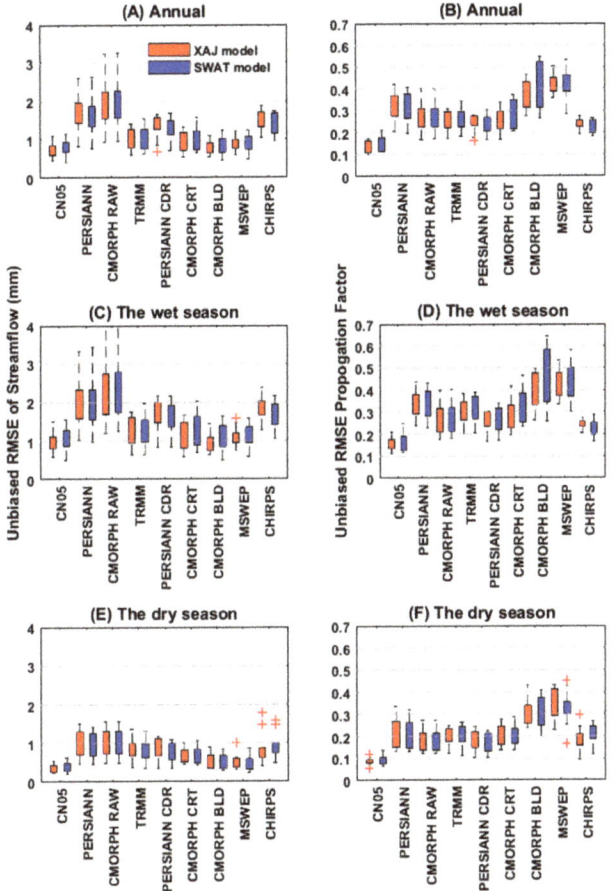

Figure 10. Unbiased RMSE of streamflow (**A,C,E**) and unbiased RMSE propagation factor (**B,D,F**) of the whole year, the wet season and the dry season for the nine gridded datasets based on SWAT (red) together with XAJ (blue) models. Each boxplot is constructed with one value of each year from 2004–2013.

Satellite-only datasets directly estimate precipitation through PMW or IR sensors [22–24]. Their worst performances among all eight satellite-based datasets reflect the defectiveness of the existent remote sensing retrievals algorithms and the necessity to blend with gauged measurement to account for their limited abilities, such as distinguishing rain particles and electromagnetic interferences from rough terrain and trees to sensors [26]. Theoretically, PMW is more accurate than VIS–IR, because the former physically links the sensors' signal to the size and phase of the hydrometeors, which is presented within the observed atmospheric column [1,24,91]. However, CMORPH RAW (mainly based primarily on PMW) performs worse than PERSIANN (mainly based on IR), which is opposite for some other regional studies in the Asian monsoon regions, such as Japan [97] and the Tibet Plateau [98]. This inconsistent result may reflect the influence of integrating methods (PMW and IR) on the performance of satellite-only datasets and the region-dependent nature of these methods.

Compared with satellite-only datasets (CMORPH RAW and PERSIANN), better performances in both precipitation and hydrological simulations are clearly achieved by their improved satellite-gauge versions (CMORPH CRT, CMORPH BLD, and PERSIANN CDR). This improvement proves the validity of the blending algorithms using gauge precipitation to enhance the precipitation estimation performances of the satellite-only datasets. Satellite-gauge CMORPH BLD outperforms all satellite-gauge datasets in both precipitation and hydrological simulations, which is mainly due to the effectiveness of using bias correction and blending algorithms by incorporating the daily precipitation gauge dataset to improve CMORPH RAW [36]. Among blended datasets, MSWEP shows comparable good performance with CMORPH BLD. The superiority of MSWEP could be mainly explained by two factors: (1) the gauged component utilized by MSWEP takes up a higher proportion (30.0% to 50.0%) in the final precipitation dataset compared to the other satellite-based datasets; and (2) the reanalysis data used in MSWEP may bring more potential information [71,99].

CN05 outperforms all satellite-based datasets in hydrological simulation. This satisfactory performance suggests that CN05 is fully able to act as the proxy of the dense-gauge precipitation dataset in the hydrological simulation in the Xiangjiang River Basin, although it could not act as the reference data to directly evaluate the statistical properties of satellite-based precipitation.

Additionally, the datasets used for model calibration would influence the hydrological performances of satellite-based datasets for the validation period, and many studies suggested recalibrating hydrological models directly using satellite-based datasets [60,100,101]. However, only the dense-gauge precipitation was used in this study to calibrate the hydrological models, and all satellite-based datasets then used the same set of optimal parameters for hydrological modeling. This is based on the assumption that the dense-gauge dataset is more accurate than the satellite-based datasets, excluding the effects of uncertainty in model parameters on hydrological simulations. Even though the satellite product-forced model performance may be degraded, when using the dense-gauge precipitation for model calibration, all satellite-based datasets used the same set of optimal parameters for hydrological modeling. In addition, one test based on the XAJ model has been conducted to prove that the calibration dataset would not change the relative hydrological performance of these satellite-based datasets, which are shown in Appendix C as Table A1 (NSE value for both calibration and validation periods) and Figure A1 (mean monthly hydrograph during 2004–2013). Therefore, it is rational to compare the performance of each satellite-based dataset. For those watersheds where the dense-gauge precipitation dataset is not available, the hydrological model may be calibrated using satellite-based datasets or other gridded datasets.

5. Conclusions

This study evaluates eight high-resolution satellite-based precipitation datasets (satellite-only: PERSIANN and CMORPH RAW, satellite-gauge: TRMM PERSIANN CDR, CMORPH CRT and CMORPH BLD and blended: MSWEP and CHIRPS) and a gauge-interpolated CN05 based on a dense-gauge dataset for hydrological modeling over a monsoon prone watershed in China. We can draw the following conclusions:

(1) All satellite-gauge and blended datasets are able to capture the seasonality of precipitation in the study region, even though biases are observed. Specifically, the satellite-gauge CMORPH BLD generally outperforms all other satellite-based datasets with the smallest detection, systematic, random errors, and most precise extreme precipitation simulation. However, satellite-only datasets perform the worst with respect to almost all the precipitation indices. Although CN05 presents the smallest systematic errors, CN05 cannot be used as the reference data to statistical analysis of the satellite-based datasets because it is missing some seasonal local precipitation and has larger random errors and a smaller R^2.

(2) There are large differences among satellite-gauge datasets in hydrological simulations. Datasets designed to provide the best instantaneous precipitation (TRMM, CMORPH CRT, CMORPH BLD, and MSWEP) perform better than those designed to achieve the most temporally

homogeneous record (PERSIANN CDR and CHIRPS). Among the four better-behaved datasets, two directly incorporating daily gauge data (CMORPH BLD and MSWEP) outperform two directly incorporating monthly gauge data (TRMM and CMORPH CRT). However, satellite-only datasets (CMORPH RAW and PERSIANN) are the least capable of simulating streamflow, which is not recommended to use in the hydrological application. CN05 outperforms all satellite-based datasets in the hydrological simulation, indicating its capability to act as reference data during the hydrological evaluation.

(3) With different model structures, XAJ and SWAT models perform differently for each satellite-based dataset, and differences in model performances also depend on seasons. Generally, the XAJ model performs better than the SWAT model in terms of random errors of streamflow simulations for both wet and dry seasons and in terms of systematic errors for the dry season. However, compared with the hydrological model uncertainties, the uncertainties from different satellite-based datasets dominate the uncertainty of hydrological simulation. In other words, the hydrological model structure does not affect the overall performance ranking of satellite-based precipitation datasets in hydrological simulations in this study.

(4) The random error from all datasets show a general decrease from precipitation to runoff with γ_{ubRMSE} being smaller than 1, but this does not hold for the systematic error with γ_{RB} varying in different datasets. In addition, the seasons and the hydrological models affect the error propagation from precipitation to streamflow for all datasets. The systematic (γ_{RB}) and random (γ_{ubRMSE}) error propagation factors of the wet season are larger than those of the dry season. The XAJ model shows a more amplified error propagation effect of the systematic errors, while the random errors are more amplified by the SWAT model.

There are still some limitations in this study. For example, the eight satellite-based datasets were compared over only one monsoon-prone watershed, and the conclusion may not be the same for other regions. In addition, the differences between using a dense-gauge dataset and satellite-based datasets to calibrate hydrological models were not fully investigated. For some data-lacking regions, the satellite-based datasets may be directly used to calibrate the hydrological models when the dense-gauge dataset is available. Therefore, in future studies, more watersheds from various climate regimes should be used to generalize the conclusions drawn from this study. In addition, the impacts of using different satellite-based datasets to calibrate the hydrological models on hydrological performances also need to be investigated.

Author Contributions: Conceptualization, J.C. and Z.L.; Data curation, Z.L., J.W. and W.Q.; Formal analysis, Z.L., L.L., J.W. and W.Q.; Funding acquisition, J.C.; Investigation, J.C., Z.L., L.L., J.W. and W.Q.; Methodology, J.C., Z.L., L.L., C.-Y.X. and J.-S.K.; Project administration, J.C.; Resources, C.-Y.X.; Supervision, J.C., C.-Y.X. and J.-S.K.; Validation, Z.L.; Visualization, J.C. and Z.L.; Writing—original draft, J.C. and Z.L.; Writing—review and editing, J.C. and Z.L. All authors have read and agreed to the published version of the manuscript.

Funding: This work was partially supported by the National Natural Science Foundation of China (Grant No. 52079093, 51779176), the Hubei Provincial Natural Science Foundation of China (Grant No. 2020CFA100), and the Overseas Expertise Introduction Project for Discipline Innovation (111 Project) (Grant No. B18037).

Acknowledgments: The authors would like to acknowledge the Water Resources Bureau of Hunan Province for providing gauged precipitation data and the China Meteorological Data Sharing System for providing temperature data and gridded datasets. The authors also thank all of the organizations for providing the satellite-based datasets, namely, the Goddard Earth Sciences Data and Information Services Center (TRMM), NOAA Climate Prediction Center (CMORPH RAW, CMORPH CRT, and CMORPH BLD), NOAA National Climatic Data Center (PERSIANN and PERSIANN CDR), Climate Hazards Group (CHIRPS) and Hylke Beck from Princeton University, the developer of MSWEP.

Conflicts of Interest: The authors declare no conflict of interest.

Appendix A

A detailed description of the satellite-based precipitation datasets used in this study.

Appendix A.1. Tropical Rainfall Measuring Mission 3B42 Dataset (TRMM)

TRMM 3B42, one of the TRMM Multi-satellite Precipitation Analysis datasets, provides 3 h precipitation estimates for the quasi-global coverage of 50°S to 50°N latitude, with coverage going back to 1998. TRMM 3B42's algorithm combines rainfall estimates from infrared and microwave satellites with gauge data and is operated in four major steps. (1) The Goddard Profiling Algorithm is used to convert and combine passive microwave data from plentiful sensors (TMI: the TRMM microwave imager, AMSU: Advanced Microwave Sounding Unit, AMSU-E: Advanced Microwave Sounding radiometer-Earth, and SSM/I: Special Sensor Microwave Imager) to 3 h precipitation estimates. (2) The infrared precipitation estimate from geostationary satellites is produced by histogram matching of the passive microwave precipitation estimate. (3) Then, microwave and infrared precipitation data are blended using infrared data to fill in the missing data where microwave data are not available. (4) The rainfall data are added up to monthly totals and blended with GPCC monthly precipitation data by using the inverse error variance weighting method to obtain the TRMM 3B43 values. The monthly sums of precipitation estimates are adjusted to make the results equal to the TRMM 3B43 monthly value. The final adjusted estimates are the TRMM 3B42 [68].

Appendix A.2. Three Climate Prediction Center Morphing Technique Datasets (CMORPH RAW, CMORPH CRT, and CMORPH BLD)

CMORPH records precipitation from 1998 to the present with a 0.07° spatial and 0.5-hourly temporal resolution [66]. CMORPH blends passive microwave-based rainfall estimates from low-orbit satellite sensors (TMI, SSM/I, AMSR-E, AMSU-B) using spatial propagation information obtained from the IR data of geostationary satellites. In this study, three different CMORPH datasets are compared: the raw satellite data (CMORPH RAW), CMORPH CRT, and CMORPH BLD. The CRT dataset is produced by blending the RAW dataset with the CPC gauge dataset and the GPCP over lands and ocean surfaces via the probability density function matching the bias correction method. An optimal interpolation method is used to combine the CRT dataset with gauge analysis to produce the BLD dataset [69].

Appendix A.3. Precipitation Estimation from Remotely Sensed Information Using Artificial Neural Networks Dataset (PERSIANN) and PERSIANN-Climate Data Record Dataset (PERSIANN CDR)

PERSIANN and PERSIANN CDR record precipitation from 2000 to the present with a spatial coverage that covers 50°S to 50°N latitude. PERSIANN utilizes infrared brightness temperature images from geostationary satellites to compute the rainfall rate estimation by the neural network function classification method [67]. Low-orbit satellite sensors (e.g., TMI, AMSR-E, AMSU-B, and SSM/I) provide passive microwave observations to update the ANN parameters. This method could narrow down the uncertainties of the relationship between cloud-top brightness temperature and precipitation estimated according to cloud properties and atmospheric conditions. Whenever independent estimates of precipitation are available, an adaptive training feature facilitates updating the network parameters [67]. Differing from PERSIANN, which is attainable in near-real time and only based on satellite measurements, PERSIANN CDR, initially designed for long-term climatic event research, is a pre-trained NCEP hourly precipitation dataset that merges the GPCP gauged dataset with its rainfall estimates [70].

Appendix A.4. Climate Hazards Group Infrared Precipitation with Station Dataset (CHIRPS)

With its period from 1981 to the present, CHIRPS is particularly designed for drought monitoring studies [102]. The latest version 2.0 dataset, issued in February 2015 is used in this study. CHIRPS is mainly produced by four steps. (1) TRMM 3B42 is used to calibrate the Cold Cloud Duration (CCD) thermal infrared data to get the 5-day CCD, which is further transformed to the fractions of the long-term mean rainfall estimation. (2) The fractions are multiplied with the monthly precipitation climatology created using rain gauge stations from the Food and Agriculture Organization (FAO) and the Global Historical Climatology Network-Monthly (GHCN) datasets in order to remove the systematic bias and produce the CHIRP dataset. (3) A modified IDW method is used to combine the CHIRP dataset with the rain gauge datasets from many archives, including the Global Telecommunication System (GTS), to obtain the CHIRPS. (4) Since the above three steps are carried out at the 5-day timescale, daily CCD data and the NOAA Climate Forecast System (CFS) atmospheric model rainfall field data are used to disaggregate the CHIRPS to daily rainfall estimates by a simple redistribution method [72].

Appendix A.5. Multi-Source Weighted-Ensemble Precipitation Dataset (MSWEP)

MSWEP is a newly raised global precipitation dataset (1979–present) with a 3-hourly temporal resolution. In order to acquire the best possible rainfall estimates, MSWEP is unique in its use of an unprecedented variety of data sources, such as gauges, atmospheric reanalysis models, and satellites. In brief, the main procedures carried out to generate the MSWEP dataset are composed of four steps. (1) The long-term mean of the MSWEP, which is mainly based on the Climate Hazards Group Precipitation Climatology (CHPclim) dataset, is bias corrected by using country-specific catch-ratio equations methods. (2) Several satellites (e.g., CMORPH, Global Satellite Mapping of Precipitation (GSMap MVK) and TRMM) and reanalysis (e.g., ERA-Interim, and the Japanese 55-year Reanalysis (JRA 55)) precipitation datasets are assessed in terms of their temporal variability to determine their potential contents in the MSWEP. (3) The long-term climatic mean is temporally downscaled in a gradual manner from monthly to daily by applying the weighted averages of precipitation anomalies from the gauge, reanalysis, and satellite datasets. (4) The long-term climatic mean is further temporally downscaled to the 3-hourly timescale to generate the final MSWEP dataset [71].

Appendix B.

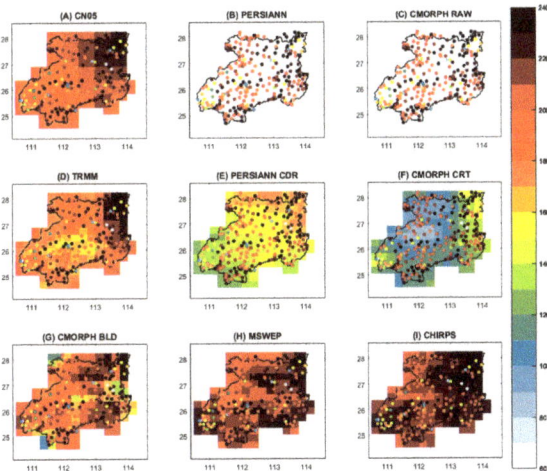

Figure A1. Winter precipitation (mm) during December–February of 2003–2013 from CN05 and eight satellite-based datasets compared with the dense-gauge dataset, which is shown as colored dots.

Appendix C.

Table A1. NSE of daily streamflow time series simulated by XAJ model calibrated by using eight satellite-based datasets, CN05, and the referenced dense-gauged dataset for calibration (2004–2010) and validation (2011–2013) periods.

Datasets	Calibrate by Dense-Gauge (NSE) 2004–2010	Validate by Dense-Gauge (NSE) 2011–2013	Calibrate by Satellite-Based Dataset (NSE)	Validate by Satellite-Based Dataset (NSE)
CN05		0.86	0.89	0.88
PERSIANN		−0.4	0.3203	−0.154
CMORPH RAW		−0.97	−0.66	−0.59
TRMM		0.73	0.79	0.72
PERSIANN CDR	0.89	0.56	0.6375	0.4783
CMORPH CRT		0.75	0.7815	0.7276
CMORPH BLD		0.84	0.8571	0.809
MSWEP		0.78	0.8517	0.799
CHIRPS		0.44	0.7019	0.42

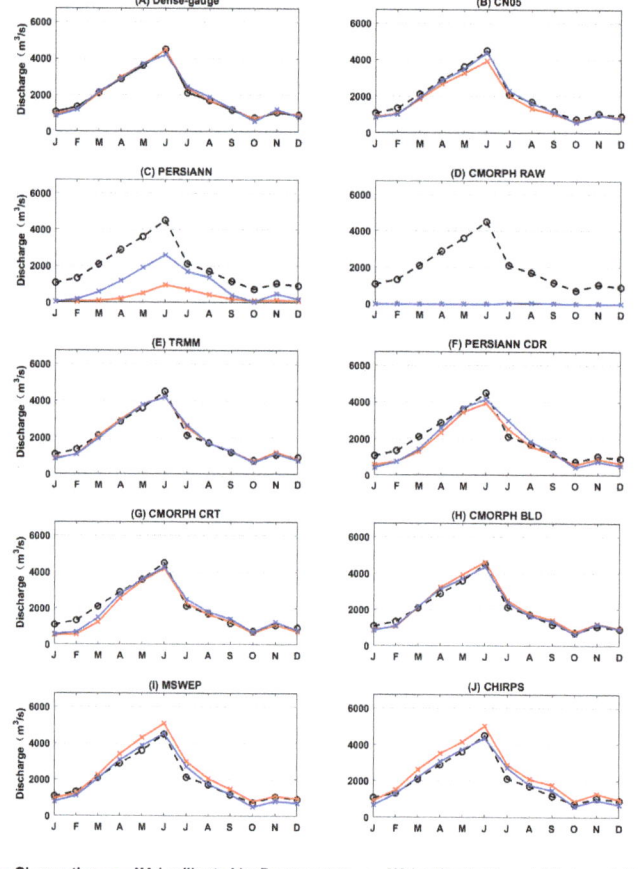

Figure A2. Mean monthly hydrographs (2004–2013) simulated by the XAJ model calibrated by using eight satellite-based datasets, CN05, and the referenced dense-gauged dataset.

References

1. Ebert, E.E.; Janowiak, J.E.; Kidd, C. Comparison of near-real-time precipitation estimates from satellite observations and numerical models. *Bull. Am. Meteorol. Soc.* **2007**, *88*, 47–64. [CrossRef]
2. Essou, G.R.; Arsenault, R.; Brissette, F.P. Comparison of climate datasets for lumped hydrological modeling over the continental United States. *J. Hydrol.* **2016**, *537*, 334–345. [CrossRef]
3. Xu, H.; Xu, C.-Y.; Chen, S.; Chen, H. Similarity and difference of global reanalysis datasets (WFD and APHRODITE) in driving lumped and distributed hydrological models in a humid region of China. *J. Hydrol.* **2016**, *542*, 343–356. [CrossRef]
4. Essou, G.R.; Sabarly, F.; Lucas-Picher, P.; Brissette, F.; Poulin, A. Can precipitation and temperature from meteorological reanalyses be used for hydrological modeling? *J. Hydrometeorol.* **2016**, *17*, 1929–1950. [CrossRef]
5. Lorenz, C.; Kunstmann, H. The hydrological cycle in three state-of-the-art reanalyses: Intercomparison and performance analysis. *J. Hydrometeorol.* **2012**, *13*, 1397–1420. [CrossRef]
6. Beck, H.E.; Vergopolan, N.; Pan, M.; Levizzani, V.; Van Dijk, A.I.; Weedon, G.P.; Brocca, L.; Pappenberger, F.; Huffman, G.J.; Wood, E.F. Global-scale evaluation of 22 precipitation datasets using gauge observations and hydrological modeling. *Hydrol. Earth Syst. Sci.* **2017**, *21*, 6201–6217. [CrossRef]
7. Blarzino, G.; Castanet, L.; Luini, L.; Capsoni, C.; Martellucci, A. 2009 3rd European Conference on Antennas and Propagation. In *Development of A New Global Rainfall Rate Model Based on ERA40, TRMM, GPCC and GPCP Products*; IEEE: New York, NY, USA, 2009; pp. 671–675.
8. Kozu, T.; Kawanishi, T.; Kuroiwa, H.; Kojima, M.; Oikawa, K.; Kumagai, H.; Okamoto, K.I.; Okumura, M.; Nakatsuka, H.; Nishikawa, K. Development of precipitation radar onboard the Tropical Rainfall Measuring Mission (TRMM) satellite. *IEEE. Trans. Geosci. Remote* **2001**, *39*, 102–116. [CrossRef]
9. Weedon, G.P.; Balsamo, G.; Bellouin, N.; Gomes, S.; Best, M.J.; Viterbo, P. The WFDEI meteorological forcing data set: WATCH Forcing Data methodology applied to ERA-Interim reanalysis data. *Water Resour. Res.* **2014**, *50*, 7505–7514. [CrossRef]
10. Yatagai, A.; Kamiguchi, K.; Arakawa, O.; Hamada, A.; Yasutomi, N.; Kitoh, A. APHRODITE: Constructing a long-term daily gridded precipitation dataset for Asia based on a dense network of rain gauges. *Bull. Am. Meteorol. Soc.* **2012**, *93*, 1401–1415. [CrossRef]
11. Rudolf, B.; Hauschild, H.; Rueth, W.; Schneider, U. Terrestrial precipitation analysis: Operational method and required density of point measurements. In *Global Precipitations and Climate Change*; Springer: Berlin, Germany, 1994; pp. 173–186.
12. New, M.; Hulme, M.; Jones, P. Representing twentieth-century space–time climate variability. Part I: Development of a 1961–90 mean monthly terrestrial climatology. *J. Climate* **1999**, *12*, 829–856. [CrossRef]
13. Hartkamp, A.D.; De Beurs, K.; Stein, A.; White, J.W. *Interpolation Techniques for Climate Variables*; CIMMYT: Mexico City, Mexico, 1999.
14. Skaugen, T.; Andersen, J. Simulated precipitation fields with variance-consistent interpolation. *Hydrolog. Sci. J.* **2010**, *55*, 676–686. [CrossRef]
15. Duan, Z.; Liu, J.; Tuo, Y.; Chiogna, G.; Disse, M. Evaluation of eight high spatial resolution gridded precipitation products in Adige Basin (Italy) at multiple temporal and spatial scales. *Sci. Total Environ.* **2016**, *573*, 1536–1553. [CrossRef] [PubMed]
16. Huang, D.Q.; Zhu, J.; Zhang, Y.C.; Huang, Y.; Kuang, X.Y. Assessment of summer monsoon precipitation derived from five reanalysis datasets over East Asia. *Q. J. R. Meteorol. Soc.* **2016**, *142*, 108–119. [CrossRef]
17. Janowiak, J.E.; Gruber, A.; Kondragunta, C.; Livezey, R.E.; Huffman, G.J. A comparison of the NCEP–NCAR reanalysis precipitation and the GPCP rain gauge–satellite combined dataset with observational error considerations. *J. Climate* **1998**, *11*, 2960–2979. [CrossRef]
18. Seyyedi, H.; Anagnostou, E.N.; Beighley, E.; McCollum, J. Hydrologic evaluation of satellite and reanalysis precipitation datasets over a mid-latitude basin. *Atmos. Res.* **2015**, *164*, 37–48. [CrossRef]
19. Kalnay, E.; Kanamitsu, M.; Kistler, R.; Collins, W.; Deaven, D.; Gandin, L.; Iredell, M.; Saha, S.; White, G.; Woollen, J. The NCEP/NCAR 40-year reanalysis project. *Bull. Am. Meteorol. Soc.* **1996**, *77*, 437–472. [CrossRef]
20. Gibson, J.K.; Kållberg, P.; Uppala, S.M.; Hernandez, A.; Nomura, A.; Serrano, E. *The ECMWF ReAnalysis (ERA) Project*; ECMWF Reanalysis Project Report Series, No. 1; ECMWF: Reading, UK, 1997; 71p.

21. Saha, S.; Moorthi, S.; Pan, H.-L.; Wu, X.; Wang, J.; Nadiga, S.; Tripp, P.; Kistler, R.; Woollen, J.; Behringer, D. The NCEP climate forecast system reanalysis. *Bull. Am. Meteorol. Soc.* **2010**, *91*, 1015–1058. [CrossRef]
22. Kidd, C.; Huffman, G. Global precipitation measurement. *Meteorol. Appl.* **2011**, *18*, 334–353. [CrossRef]
23. Maggioni, V.; Meyers, P.C.; Robinson, M.D. A review of merged high-resolution satellite precipitation product accuracy during the Tropical Rainfall Measuring Mission (TRMM) era. *J. Hydrometeorol.* **2016**, *17*, 1101–1117. [CrossRef]
24. Mei, Y.; Anagnostou, E.N.; Nikolopoulos, E.I.; Borga, M. Error analysis of satellite precipitation products in mountainous basins. *J. Hydrometeorol.* **2014**, *15*, 1778–1793. [CrossRef]
25. Sapiano, M.; Arkin, P. An intercomparison and validation of high-resolution satellite precipitation estimates with 3-hourly gauge data. *J. Hydrometeorol.* **2009**, *10*, 149–166. [CrossRef]
26. Mizukami, N.; Smith, M.B. Analysis of inconsistencies in multi-year gridded quantitative precipitation estimate over complex terrain and its impact on hydrologic modeling. *J. Hydrol.* **2012**, *428*, 129–141. [CrossRef]
27. Huffman, G.J.; Adler, R.F.; Arkin, P.; Chang, A.; Ferraro, R.; Gruber, A.; Janowiak, J.; McNab, A.; Rudolf, B.; Schneider, U. The global precipitation climatology project (GPCP) combined precipitation dataset. *Bull. Am. Meteorol. Soc.* **1997**, *78*, 5–20. [CrossRef]
28. Hsu, K.-l.; Gao, X.; Sorooshian, S.; Gupta, H.V. Precipitation estimation from remotely sensed information using artificial neural networks. *J. Appl. Meteorol.* **1997**, *36*, 1176–1190. [CrossRef]
29. Huffman, G.J.; Bolvin, D.T.; Nelkin, E.J.; Wolff, D.B.; Adler, R.F.; Gu, G.; Hong, Y.; Bowman, K.P.; Stocker, E.F. The TRMM multisatellite precipitation analysis (TMPA): Quasi-global, multiyear, combined-sensor precipitation estimates at fine scales. *J. Hydrometeorol.* **2007**, *8*, 38–55. [CrossRef]
30. Hou, A.Y.; Kakar, R.K.; Neeck, S.; Azarbarzin, A.A.; Kummerow, C.D.; Kojima, M.; Oki, R.; Nakamura, K.; Iguchi, T. The global precipitation measurement mission. *Bull. Am. Meteorol. Soc.* **2014**, *95*, 701–722. [CrossRef]
31. Behrangi, A.; Lebsock, M.; Wong, S.; Lambrigtsen, B. On the quantification of oceanic rainfall using spaceborne sensors. *J. Geophys. Res. Atmos.* **2012**, *117*. [CrossRef]
32. Tapiador, F.; Navarro, A.; Levizzani, V.; García-Ortega, E.; Huffman, G.; Kidd, C.; Kucera, P.; Kummerow, C.; Masunaga, H.; Petersen, W. Global precipitation measurements for validating climate models. *Atmos. Res.* **2017**, *197*, 1–20. [CrossRef]
33. Sun, Q.; Miao, C.; Duan, Q.; Ashouri, H.; Sorooshian, S.; Hsu, K.L. A review of global precipitation data sets: Data sources, estimation, and intercomparisons. *Rev. Geophys.* **2018**, *56*, 79–107. [CrossRef]
34. Voisin, N.; Wood, A.W.; Lettenmaier, D.P. Evaluation of precipitation products for global hydrological prediction. *J. Hydrometeorol.* **2008**, *9*, 388–407. [CrossRef]
35. Maggioni, V.; Massari, C. On the performance of satellite precipitation products in riverine flood modeling: A review. *J. Hydrol.* **2018**, *558*, 214–224. [CrossRef]
36. Behrangi, A.; Khakbaz, B.; Jaw, T.C.; AghaKouchak, A.; Hsu, K.; Sorooshian, S. Hydrologic evaluation of satellite precipitation products over a mid-size basin. *J. Hydrol.* **2011**, *397*, 225–237. [CrossRef]
37. Bodian, A.; Dezetter, A.; Deme, A.; Diop, L. Hydrological evaluation of TRMM rainfall over the upper Senegal River basin. *Hydrology* **2016**, *3*, 15. [CrossRef]
38. Dile, Y.T.; Srinivasan, R. Evaluation of CFSR climate data for hydrologic prediction in data-scarce watersheds: An application in the Blue Nile River Basin. *J. Am. Water Resour. Assoc.* **2014**, *50*, 1226–1241. [CrossRef]
39. Fujihara, Y.; Yamamoto, Y.; Tsujimoto, Y.; Sakagami, J.-I. Discharge simulation in a data-scarce basin using reanalysis and global precipitation data: A case study of the White Volta Basin. *J. Water Resour. Prot.* **2014**, *6*, 1316. [CrossRef]
40. Getirana, A.C.; Espinoza, J.; Ronchail, J.; Rotunno Filho, O. Assessment of different precipitation datasets and their impacts on the water balance of the Negro River basin. *J. Hydrol.* **2011**, *404*, 304–322. [CrossRef]
41. Poméon, T.; Jackisch, D.; Diekkrüger, B. Evaluating the performance of remotely sensed and reanalysed precipitation data over West Africa using HBV light. *J. Hydrol.* **2017**, *547*, 222–235. [CrossRef]
42. Su, J.; Lü, H.; Wang, J.; Sadeghi, A.M.; Zhu, Y. Evaluating the applicability of four latest satellite–gauge combined precipitation estimates for extreme precipitation and streamflow predictions over the upper Yellow River basins in China. *Remote Sens.* **2017**, *9*, 1176. [CrossRef]
43. Tuo, Y.; Duan, Z.; Disse, M.; Chiogna, G. Evaluation of precipitation input for SWAT modeling in Alpine catchment: A case study in the Adige river basin (Italy). *Sci. Total Environ.* **2016**, *573*, 66–82. [CrossRef]

44. Zhu, Q.; Hsu, K.-l.; Xu, Y.P.; Yang, T. Evaluation of a new satellite-based precipitation data set for climate studies in the Xiang River basin, southern China. *Int. J. Climatol.* **2017**, *37*, 4561–4575. [CrossRef]
45. Nikolopoulos, E.I.; Anagnostou, E.N.; Hossain, F.; Gebremichael, M.; Borga, M. Understanding the scale relationships of uncertainty propagation of satellite rainfall through a distributed hydrologic model. *J. Hydrometeorol.* **2010**, *11*, 520–532. [CrossRef]
46. Duncan, J.M.; Biggs, E.M. Assessing the accuracy and applied use of satellite-derived precipitation estimates over Nepal. *Appl. Geogr.* **2012**, *34*, 626–638. [CrossRef]
47. Khan, S.I.; Hong, Y.; Gourley, J.J.; Khattak, M.U.K.; Yong, B.; Vergara, H.J. Evaluation of three high-resolution satellite precipitation estimates: Potential for monsoon monitoring over Pakistan. *Adv. Space Res.* **2014**, *54*, 670–684. [CrossRef]
48. Mao, J.; Wu, G. Diurnal variations of summer precipitation over the Asian monsoon region as revealed by TRMM satellite data. *Sci. China Earth Sci.* **2012**, *55*, 554–566. [CrossRef]
49. Mou, T.; Ab, I.; Duan, Z.; Arthur, C.; Vincent, C. Evaluation of Six High-Resolution Satellite and Ground-Based Precipitation Products over Malaysia. *Remote Sens.* **2015**, *7*, 1504–1528.
50. Prakash, S.; Mitra, A.K.; AghaKouchak, A.; Pai, D. Error characterization of TRMM Multisatellite Precipitation Analysis (TMPA-3B42) products over India for different seasons. *J. Hydrol.* **2015**, *529*, 1302–1312. [CrossRef]
51. Prakash, S.; Sathiyamoorthy, V.; Mahesh, C.; Gairola, R. An evaluation of high-resolution multisatellite rainfall products over the Indian monsoon region. *Int. J. Remote Sens.* **2014**, *35*, 3018–3035. [CrossRef]
52. Sunilkumar, K.; Narayana Rao, T.; Saikranthi, K.; Purnachandra Rao, M. Comprehensive evaluation of multisatellite precipitation estimates over India using gridded rainfall data. *J. Geophys. Res. Atmos.* **2015**, *120*, 8987–9005. [CrossRef]
53. Bajracharya, S.; Shrestha, M.; Shrestha, A. Assessment of high-resolution satellite rainfall estimation products in a streamflow model for flood prediction in the Bagmati basin, Nepal. *J. Flood Risk Manag.* **2017**, *10*, 5–16. [CrossRef]
54. Jiang, S.; Ren, L.; Hong, Y.; Yang, X.; Ma, M.; Zhang, Y.; Yuan, F. Improvement of multi-satellite real-time precipitation products for ensemble streamflow simulation in a middle latitude basin in South China. *Water Resour. Manag.* **2014**, *28*, 2259–2278. [CrossRef]
55. Kim, J.P.; Jung, I.W.; Park, K.W.; Yoon, S.K.; Lee, D. Hydrological utility and uncertainty of multi-satellite precipitation products in the mountainous region of South Korea. *Remote Sens.* **2016**, *8*, 608.
56. Lauri, H.; Räsänen, T.; Kummu, M. Using reanalysis and remotely sensed temperature and precipitation data for hydrological modeling in monsoon climate: Mekong River case study. *J. Hydrometeorol.* **2014**, *15*, 1532–1545.
57. Li, L.; Xu, C.-Y.; Zhang, Z.; Jain, S.K. Validation of a new meteorological forcing data in analysis of spatial and temporal variability of precipitation in India. *Stoch. Environ. Res. Risk Assess.* **2014**, *28*, 239–252.
58. Tong, K.; Su, F.; Yang, D.; Hao, Z. Evaluation of satellite precipitation retrievals and their potential utilities in hydrologic modeling over the Tibetan Plateau. *J. Hydrol.* **2014**, *519*, 423–437.
59. Wang, S.; Liu, S.; Mo, X.; Peng, B.; Qiu, J.; Li, M.; Liu, C.; Wang, Z.; Bauer-Gottwein, P. Evaluation of remotely sensed precipitation and its performance for streamflow simulations in basins of the southeast Tibetan Plateau. *J. Hydrometeorol.* **2015**, *16*, 2577–2594.
60. Xue, X.; Hong, Y.; Limaye, A.S.; Gourley, J.J.; Huffman, G.J.; Khan, S.I.; Dorji, C.; Chen, S. Statistical and hydrological evaluation of TRMM-based Multi-satellite Precipitation Analysis over the Wangchu Basin of Bhutan: Are the latest satellite precipitation products 3B42V7 ready for use in ungauged basins? *J. Hydrol.* **2013**, *499*, 91–99.
61. Gao, X.-J. A gridded daily observation dataset over China region and comparison with the other datasets. *Diqiu Wuli Xuebao* **2013**, *56*, 1102–1111.
62. Sun, Q.; Miao, C.; Duan, Q.; Kong, D.; Ye, A.; Di, Z.; Gong, W. Would the 'real' observed dataset stand up? A critical examination of eight observed gridded climate datasets for China. *Environ. Res. Lett.* **2014**, *9*, 015001.
63. Yang, F.; Lu, H.; Yang, K.; He, J.; Wang, W.; Wright, J.S.; Li, C.; Han, M.; Li, Y. Evaluation of multiple forcing data sets for precipitation and shortwave radiation over major land areas of China. *Hydrol. Earth Syst. Sci.* **2017**, *21*. [CrossRef]
64. Zhou, B.; Xu, Y.; Wu, J.; Dong, S.; Shi, Y. Changes in temperature and precipitation extreme indices over China: Analysis of a high-resolution grid dataset. *Int. J. Climatol.* **2016**, *36*, 1051–1066.

65. Ma, C.; Pan, S.; Wang, G.; Liao, Y.; Xu, Y.-P. Changes in precipitation and temperature in Xiangjiang River Basin, China. *Theor. Appl. Climatol.* **2016**, *123*, 859–871. [CrossRef]
66. Joyce, R.J.; Janowiak, J.E.; Arkin, P.A.; Xie, P. CMORPH: A method that produces global precipitation estimates from passive microwave and infrared data at high spatial and temporal resolution. *J. Hydrometeorol.* **2004**, *5*, 487–503. [CrossRef]
67. Sorooshian, S.; Hsu, K.-L.; Gao, X.; Gupta, H.V.; Imam, B.; Braithwaite, D. Evaluation of PERSIANN system satellite-based estimates of tropical rainfall. *Bull. Am. Meteorol. Soc.* **2000**, *81*, 2035–2046. [CrossRef]
68. Huffman, G.J.; Adler, R.F.; Bolvin, D.T.; Nelkin, E.J. The TRMM multi-satellite precipitation analysis (TMPA). In *Satellite Rainfall Applications for Surface Hydrology*; Springer: Berlin, Germany, 2010; pp. 3–22.
69. Xie, P.; Xiong, A.Y. A conceptual model for constructing high-resolution gauge-satellite merged precipitation analyses. *J. Geophys. Res. Atmos.* **2011**, *116*. [CrossRef]
70. Ashouri, H.; Hsu, K.-L.; Sorooshian, S.; Braithwaite, D.K.; Knapp, K.R.; Cecil, L.D.; Nelson, B.R.; Prat, O.P. PERSIANN-CDR: Daily precipitation climate data record from multisatellite observations for hydrological and climate studies. *Bull. Am. Meteorol. Soc.* **2015**, *96*, 69–83. [CrossRef]
71. Beck, H.E.; Van Dijk, A.I.; Levizzani, V.; Schellekens, J.; Gonzalez Miralles, D.; Martens, B.; De Roo, A. MSWEP: 3-hourly 0.25 global gridded precipitation (1979–2015) by merging gauge, satellite, and reanalysis data. *Hydrol. Earth Syst. Sci.* **2017**, *21*, 589–615. [CrossRef]
72. Peterson, P. The Climate Hazards Group InfraRed Precipitation with Stations (CHIRPS) v2.0 Dataset: 35 year Quasi-Global Precipitation Estimates for Drought Monitoring. *Sci. Data* **2014**, *2*, 1–21.
73. Yang, J. The thin plate spline robust point matching (TPS-RPM) algorithm: A revisit. *Pattern Recogn. Lett.* **2011**, *32*, 910–918. [CrossRef]
74. Ruelland, D.; Ardoin-Bardin, S.; Billen, G.; Servat, E. Sensitivity of a lumped and semi-distributed hydrological model to several methods of rainfall interpolation on a large basin in West Africa. *J. Hydrol.* **2008**, *361*, 96–117. [CrossRef]
75. Luo, Q.; Li, Y.; Wang, K.; Wu, J. Application of the SWAT model to the Xiangjiang river watershed in subtropical central China. *Water Sci. Technol.* **2013**, *67*, 2110–2116. [CrossRef]
76. Xu, H.; Xu, C.-Y.; Chen, H.; Zhang, Z.; Li, L. Assessing the influence of rain gauge density and distribution on hydrological model performance in a humid region of China. *J. Hydrol.* **2013**, *505*, 1–12. [CrossRef]
77. Zhao, R. The Xinanjiang model applied in China. *J. Hydrol.* **1992**, *135*, 371–381.
78. Zhao, R.; Liu, X. *The Xinanjiang Model, Computer Models of Watershed Hydrology*; Singh, V.P., Ed.; Water Resources Publications: California City, CA, USA, 1995; pp. 215–232.
79. Wang, W.-C.; Cheng, C.-T.; Chau, K.-W.; Xu, D.-M. Calibration of Xinanjiang model parameters using hybrid genetic algorithm based fuzzy optimal model. *J. Hydroinform.* **2012**, *14*, 784–799. [CrossRef]
80. Yan, R.; Huang, J.; Wang, Y.; Gao, J.; Qi, L. Modeling the combined impact of future climate and land use changes on streamflow of Xinjiang Basin, China. *Hydrol. Res.* **2016**, *47*, 356–372. [CrossRef]
81. Zeng, Q.; Chen, H.; Xu, C.-Y.; Jie, M.-X.; Hou, Y.-K. Feasibility and uncertainty of using conceptual rainfall-runoff models in design flood estimation. *Hydrol. Res.* **2016**, *47*, 701–717. [CrossRef]
82. Arnold, J.G.; Srinivasan, R.; Muttiah, R.S.; Williams, J.R. Large area hydrologic modeling and assessment part I: Model development 1. *J. Am. Water Resour. Assoc.* **1998**, *34*, 73–89. [CrossRef]
83. Awan, U.K.; Liaqat, U.W.; Choi, M.; Ismaeel, A. A SWAT modeling approach to assess the impact of climate change on consumptive water use in Lower Chenab Canal area of Indus basin. *Hydrol. Res.* **2016**, *47*, 1025–1037. [CrossRef]
84. Aouissi, J.; Benabdallah, S.; Chabaane, Z.L.; Cudennec, C. Evaluation of potential evapotranspiration assessment methods for hydrological modelling with SWAT—Application in data-scarce rural Tunisia. *Agric. Water Manag.* **2016**, *174*, 39–51. [CrossRef]
85. Jha, M.; Pan, Z.; Takle, E.S.; Gu, R. Impacts of climate change on streamflow in the Upper Mississippi River Basin: A regional climate model perspective. *J. Geophys. Res. Atmos.* **2004**, *109*. [CrossRef]
86. Duan, Q.; Sorooshian, S.; Gupta, V. Effective and efficient global optimization for conceptual rainfall-runoff models. *Water Resour. Res.* **1992**, *28*, 1015–1031. [CrossRef]
87. Abbaspour, K.C.; Johnson, C.; Van Genuchten, M.T. Estimating uncertain flow and transport parameters using a sequential uncertainty fitting procedure. *Vadose Zone J.* **2004**, *3*, 1340–1352.

88. Wang, J.; Chen, H.; Xu, C.-Y.; Zeng, Q.; Wang, Q.; Kim, J.-S.; Chen, J.; Guo, S. Tracking the error sources of spatiotemporal differences in TRMM accuracy using error decomposition method. *Hydrol. Res.* **2018**, *49*, 1960–1976.
89. Kidd, C.; Kniveton, D.R.; Todd, M.C.; Bellerby, T.J. Satellite rainfall estimation using combined passive microwave and infrared algorithms. *J. Hydrometeorol.* **2003**, *4*, 1088–1104. [CrossRef]
90. Bell, T.L.; Kundu, P.K. Dependence of satellite sampling error on monthly averaged rain rates: Comparison of simple models and recent studies. *J. Climate* **2000**, *13*, 449–462.
91. Bennartz, R.; Petty, G.W. The sensitivity of microwave remote sensing observations of precipitation to ice particle size distributions. *J. Appl. Meteorol.* **2001**, *40*, 345–364. [CrossRef]
92. Chang, A.T.; Chiu, L.S. Nonsystematic errors of monthly oceanic rainfall derived from SSM/I. *Mon. Weather. Rev.* **1999**, *127*, 1630–1638. [CrossRef]
93. Kummerow, C. Beamfilling errors in passive microwave rainfall retrievals. *J. Appl. Meteorol.* **1998**, *37*, 356–370. [CrossRef]
94. Nash, J.E.; Sutcliffe, J.V. River flow forecasting through conceptual models part I—A discussion of principles. *J. Hydrol.* **1970**, *10*, 282–290. [CrossRef]
95. Epstein, E.S. A spectral climatology. *J. Climate* **1988**, *1*, 88–107.
96. Narapusetty, B.; Timothy, D.; Michael, K.T. Optimal estimation of the climatological mean. *J. Climate* **2009**, *22*, 4845–4859.
97. Kubota, T.; Ushio, T.; Shige, S.; Kida, S.; Kachi, M.; Okamoto, K.I. Verification of high-resolution satellite-based rainfall estimates around Japan using a gauge-calibrated ground-radar dataset. *J. Meteorol. Soc. Jpn. Ser. II* **2009**, *87*, 203–222. [CrossRef]
98. Gao, Y.; Liu, M. Evaluation of high-resolution satellite precipitation products using rain gauge observations over the Tibetan Plateau. *Hydrol. Earth Syst. Sci.* **2013**, *17*, 837–849. [CrossRef]
99. Alijanian, M.; Rakhshandehroo, G.R.; Mishra, A.K.; Dehghani, M. Evaluation of satellite rainfall climatology using CMORPH, PERSIANN-CDR, PERSIANN, TRMM, MSWEP over Iran. *Int. J. Climatol.* **2017**, *37*, 4896–4914. [CrossRef]
100. Bitew, M.M.; Gebremichael, M. Assessment of satellite rainfall products for streamflow simulation in medium watersheds of the Ethiopian highlands. *Hydrol. Earth Syst. Sci.* **2011**, *15*, 1147. [CrossRef]
101. Yong, B.; Hong, Y.; Ren, L.L.; Gourley, J.J.; Huffman, G.J.; Chen, X.; Wang, W.; Khan, S.I. Assessment of evolving TRMM-based multisatellite real-time precipitation estimation methods and their impacts on hydrologic prediction in a high latitude basin. *J. Geophys. Res. Atmos.* **2012**, *117*. [CrossRef]
102. Funk, C.; Peterson, P.; Landsfeld, M.; Pedreros, D.; Verdin, J.; Shukla, S.; Husak, G.; Rowland, J.; Harrison, L.; Hoell, A. The climate hazards infrared precipitation with stations—A new environmental record for monitoring extremes. *Sci. Data* **2015**, *2*, 1–21. [CrossRef]

Publisher's Note: MDPI stays neutral with regard to jurisdictional claims in published maps and institutional affiliations.

© 2020 by the authors. Licensee MDPI, Basel, Switzerland. This article is an open access article distributed under the terms and conditions of the Creative Commons Attribution (CC BY) license (http://creativecommons.org/licenses/by/4.0/).

Article

Spatial Downscaling of MODIS Chlorophyll-a with Genetic Programming in South Korea

Hamid Mohebzadeh, Junho Yeom and Taesam Lee *

Department of Civil Engineering, ERI, Gyeongsang National University, 501 Jinju-daero, Jinju, Gyeongnam 52828, Korea; hamidmohebzadeh@gnu.ac.kr (H.M.); junho.yeom@gnu.ac.kr (J.Y.)
* Correspondence: tae3lee@gnu.ac.kr

Received: 20 March 2020; Accepted: 28 April 2020; Published: 30 April 2020

Abstract: Chlorophyll-a (Chl-a) is one of the major indicators for water quality assessment and recent developments in ocean color remote sensing have greatly improved the ability to monitor Chl-a on a global scale. The coarse spatial resolution is one of the major limitations for most ocean color sensors including Moderate Resolution Imaging Spectroradiometer (MODIS), especially in monitoring the Chl-a concentrations in coastal regions. To improve its spatial resolution, downscaling techniques have been suggested with polynomial regression models. Nevertheless, polynomial regression has some restrictions, including sensitivity to outliers and fixed mathematical forms. Therefore, the current study applied genetic programming (GP) for downscaling Chl-a. The proposed GP model in the current study was compared with multiple polynomial regression (MPR) to different degrees (2nd-, 3rd-, and 4th-degree) to illustrate their performances for downscaling MODIS Chl-a. The obtained results indicate that GP with R^2 = 0.927 and RMSE = 0.1642 on the winter day and R^2 = 0.763 and RMSE = 0.5274 on the summer day provides higher accuracy on both winter and summer days than all the applied MPR models because the GP model can automatically produce appropriate mathematical equations without any restrictions. In addition, the GP model is the least sensitive model to the changes in the input parameters. The improved downscaling data provide better information to monitor the status of oceanic and coastal marine ecosystems that are also critical for fisheries and fishing farming.

Keywords: spatial downscaling; MODIS chlorophyll-a; sentinel-2A MSI; multiple polynomial regression; genetic programming

1. Introduction

Coastal marine ecosystems are the most important habitats for species that live in the world's most productive ecosystems, such as fish and marine mammals [1]. The influences of the proximity to land, large quantities of nutrients delivered via streams, and sewage discharge lead to increased susceptibility of these ecosystems to rapid changes in water quality through both anthropogenic and natural mechanisms. Therefore, it is essential to monitor water quality in coastal ecosystems to mitigate the adverse impacts of human-related activities in these environments [1,2].

A phytoplankton cell is a planktonic photosynthesizing organism [3], and phytoplankton biomass can serve an index to provide information about marine ecosystem health. Coastal ecosystems throughout the world are affected by the fast growth of the phytoplankton population, often resulting from water column stratification or increases in nutrients [2]. Harmful algal blooms, like dinoflagellate, Gymnodinium breve (commonly referred to as "red tide") produce neurotoxins such as saxitoxin and gonyautotoxin that cause water quality degradation, which have considerable consequences for marine environments such as fish death [4–7]. The chlorophyll-a (Chl-a) concentration has been recognized as a direct indicator of phytoplankton biomass because all phytoplanktons contain Chl-a and high

Chl-a concentrations show more desirable environmental conditions for phytoplankton growth [3]. Therefore, by monitoring changes in the Chl-a concentration distribution, long-term trends in the water quality of coastal and oceanic systems can be assessed to a point where the negative effects can be mitigated [2,8,9]. However, traditional techniques such as in situ field sampling, moored instruments, drifting instruments, and fluorometry used for Chl-a measurement are expensive laboratory-based instruments and have some spatiotemporal limitations.

Over the last two decades, there has been an increase in remote sensing applications as a substitute for traditional techniques for near-real-time measurements of global phytoplankton biomass, including both qualitative and quantitative estimates [10,11]. However, there are two major challenges associated with extracting information from remote sensing data: 1) the sheer amount of data, and 2) variable precision and continuity among remote sensing-derived products. [12]. To solve such issues, many techniques have been developed, such as reflectance-based classification algorithms [13], spectral band ratios [14–16], spectral band-difference algorithms [17–19], bio-optical models [20,21], and analytical techniques [22,23].

The retrieval of Chl-a concentrations in coastal areas by the abovementioned techniques is performed using coarse spatial resolution ocean color sensors such as Moderate Resolution Imaging Spectroradiometer (MODIS), Coastal Zone Color Scanner (CZCS), MEdium Resolution Imaging Spectrometer (MERIS), and Sea-viewing Wide Field-of-view Sensor (SeaWiFS). Although the high temporal resolution of these sensors (e.g., 1–2 days for MODIS, 3 days for MERIS, and 1 day for SeaWiFS) makes them suitable for continuous monitoring, their spatial resolution (e.g., 4 km for MODIS, 300 m for MERIS, and 1.1 km for SeaWiFS) is not satisfactory due to their orbital characteristics and technical configurations.

Recent studies have introduced spatial downscaling algorithms as an alternative solution to the coarse spatial resolution of ocean color sensors. Spatial downscaling has been widely used for downscaling coarse spatial resolution data by utilizing the high-resolution remote sensing reflectance measurements, for instance for land surface temperature [24–26], precipitation [27–30], and soil moisture [31,32]. Fu, et al. [33] combined coarse spatial resolution MODIS Chl-a measurements with high spatial resolution Landsat 8 OLI band combinations using a polynomial regression model (fourth-order polynomial regression) to downscale MODIS Chl-a maps from 4 km to 30 m spatial resolution. However, polynomial regression models have some restrictions, such as sensitivity to outliers and the use of fixed mathematical forms to define the relationship between the predictor and predictand variables. Machine learning (ML) approaches have been suggested to deal with these restrictions and have received increasing attention for downscaling studies as powerful alternative tools, but only limited applications for downscaling of Chl-a [34–36].

Among ML models, genetic programming (GP) have recently received much attention in a number of fields including water resource management studies [37]. The idea behind the GP was inspired by biological evolution that makes it a collection of techniques for finding the best solution in the space of possible solutions. This unique feature of GP made it a suitable technique for various water resource management applications, including ocean engineering and hydrology, hydrological forecasts, and groundwater modeling [37]. Therefore, the current study assessed the accuracy of GP for Chl-a downscaling and compared its results with the results of three multiple polynomial regression (MPR) models, including second-order (2nd), third-order (3rd), and fourth-order (4th) polynomials. The developed models were utilized for Chl-a downscaling over the western coast of South Korea.

2. Study Area

The study area was part of the Korean West Sea, which is located in the eastern part of the Yellow Sea (35°15′ − 36°30′ N, 125°45′ − 126°45′ E; area of 10,705 km^2) (Figure 1). The Yellow Sea is a shallow marine ecosystem with the average and maximum water depths of 44 m and 103 m, respectively [38]. There is clear seasonality in sea surface temperature (SST) over the Yellow Sea, where January is the coldest month with an average SST of 4–7 °C and July is the warmest, with an average SST of

26–27 °C [39]. There are a total of 339 fish species in the Korean West Sea [40]. Over the past few years, some fish species, such as small yellow croaker, hairtail, large yellow croaker, and flatfish have exhibited continuous declines due to overharvesting, degraded marine ecosystem quality, and several unknown factors [41].

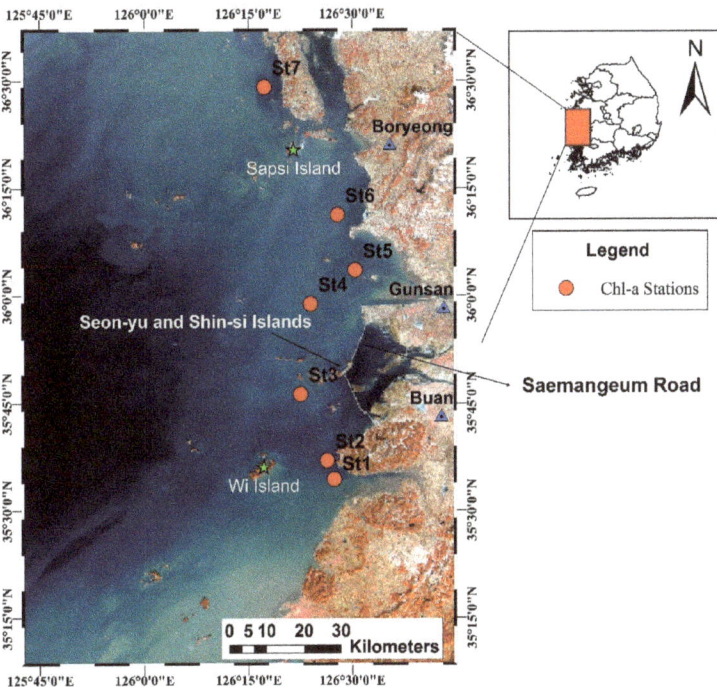

Figure 1. Location of the study region in the Korean West Sea and Chl-a sampling stations.

The increase in the level of the eutrophication, as a result of human activities such as dense agricultural practices along the coastal area, is another reason for environmental pollution over the study area that have significant negative effects on marine ecosystems, such as fish death, and a loss of important protein for the people dependent upon them [1]. Furthermore, in the Yellow Sea waters, there are other constituents than phytoplankton, such as inorganic particles and dissolved organic matter that are the major obstacle for simple empirical algorithms to determine the statistical relationship between Chl-a concentration and spectral bands [42]. Additionally, a very limited number of studies have attempted to investigate the optical properties of the Yellow Sea, such as phytoplankton, from the ocean color images. As a result, monitoring the Chl-a concentration, as an important index to evaluate the extent of eutrophication, at fine resolution is a crucial task in this area.

3. Materials and Methods

The main objective of this research was to develop an approach for MODIS Chl-a downscaling and produce Chl-a concentration maps for complex coastal regions. Figure 2 displays the detailed explanation of the procedure used. The downscaling approach was accomplished in four steps: (1) remote sensing data, including MODIS Chl-a at 4 km (defined as Y_{4k}) and S-2A at 10 m (defined as X_{10}), were acquired, and S-2A data were resampled to 4 km MODIS resolution (denoted as X_{4k}); (2) the most important S-2A band combinations (X_{4k}) were chosen by utilizing the support vector machine recursive feature elimination (SVM-RFE) method; (3) MODIS Chl-a downscaling from 4 km to 10 m was performed by regressing X_{4k} to Y_{4k}, calculating the residual at 4 km (ε_{4k}), and adding the interpolated residual (ε_{10}) to the estimated fine-resolution Chl-a ($\hat{Y}_{10}$); (4) the obtained downscaled maps were compared with visual comparison, validated with in situ data, and all the applied methods were assessed using sensitivity analysis. A complete explanation of the aforementioned steps is described in Section 3.1, Section 3.2, Section 3.3, Section 3.4.

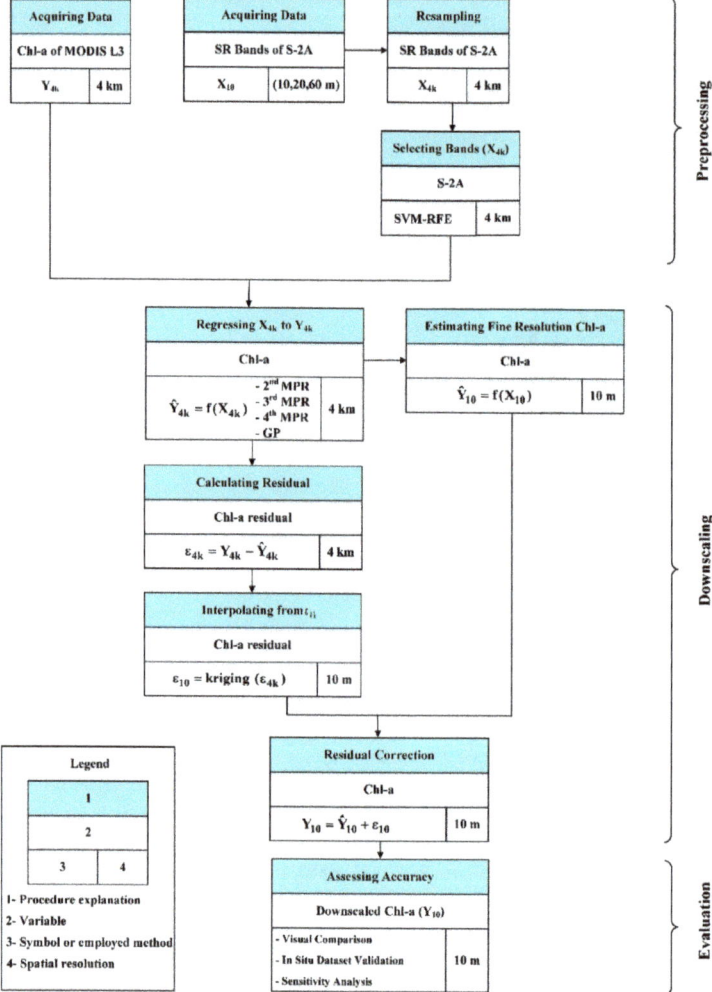

Figure 2. Downscaling workflow. Note that the goal of the present research is to downscale Moderate Resolution Imaging Spectroradiometer (MODIS) Chl-a from coarse resolution (4 km) to high resolution (10 m).

3.1. Data Description

In the current study, a downscaling approach was proposed based on GP and MPR techniques to produce high-resolution MODIS Chl-a data by relating coarse-resolution MODIS Chl-a data to high-resolution Sentinel-2A MSI (S-2A) data. There are some challenges associated with downscaling Chl-a: 1) MODIS Chl-a and S-2A measurements have different revisit times of 8 and 10 days, respectively; 2) in situ data used for validation of the downscaling model are irregularly distributed in space and rarely accessible; 3) to obtain reasonable results for downscaling Chl-a, a remote sensing image should have cloud coverage less than 10%. It has been reported that sea fog is frequently observed around the Korean peninsula with a maximum occurrence in the West Sea in summer with a mean frequency of 5.3 days in July [43]. The frequency is less than 1 day during fall and winter. Therefore, the above-mentioned challenges imposed some limitations for selection of images on every season. Hence, two MODIS and S-2A Level-1C (hereafter S-2A) datasets were acquired on February 2 (winter) and August 4 (summer) in 2016. All images had less than 10% cloud coverage and were almost concurrent with in situ measurements (±5 days).

The Chl-a concentration maps used in the current study were used from MODIS Chl-a concentration standard mapped images (SMIs) with a 4 km resolution. The SMIs were created from bands 13 (0.653 μm),14 (0.681 μm), and 15 (0.750 μm) of the MODIS sensor (hereafter referred to as MODIS Chl-a). Due to strong fluorescence signals from Chl-a at these spectral bands, these bands were suitable for the detection of the Chl-a concentration [44,45]. For retrieval of Chl-a concentration, the standard OC3/OC4 (OCx) band ratio algorithm combined with color index (CI) introduced by Reference [46] was implemented. The MODIS Chl-a data were obtained from the Giovanni website (https://giovanni.gsfc.nasa.gov/giovanni/).

S-2A band images (Table 1) for high-resolution data were acquired from the Copernicus Open Access Hub (https://scihub.copernicus.eu/). S-2A products include 13 bands with the top-of-atmosphere (TOA) Reflectance. All bands were processed by radiometric and geometric correction using a UTM/WGS84 projection. Recent studies have reported that S-2A is more powerful than Landsat 8 OLI in distinguishing areas affected by algal blooms because S-2A has three near-infrared (NIR) bands and Landsat 8 OLI has only one NIR band [47–51]. The three additional NIR bands of S-2A make it feasible to develop more suitable algorithms for the retrieval of the Chl-a concentration in intense bloom conditions [52].

Table 1. Sentinel-2A MSI spectral band characteristics.

Band Name	Central Wavelength (μm)	Spatial Resolution (m)
B1-Coastal aerosol	0.443	60
B2-Blue	0.490	10
B3-Green	0.560	10
B4-Red	0.665	10
B5-Vegetation red edge	0.705	20
B6-Vegetation red edge	0.740	20
B7-Vegetation red edge	0.783	20
B8-NIR	0.842	10
B8A-Narrow NIR	0.865	20
B9-Water vapor	0.940	60
B10-SWIR-Cirrus	1.375	60
B11-SWIR	1.610	20
B12-SWIR	2.190	20

The in situ Chl-a measurements used in the analysis were obtained for seven stations from the Korea Institute of Ocean Science and Technology (KIOST) website (http://joiss.kr). The locations of the stations are shown in Figure 1. Multidepth sampling was used to collect water samples from the surface layer (0–15 cm depth). A fairly large water sample was collected and the sample was filtered to concentrate the chlorophyll-containing organisms followed by mechanical rupturing of the collected cells, and extraction of the chlorophyll from the disrupted cells into the organic solvent acetone. The extract was then analyzed by a spectrophotometric method using fluorescence. Therefore, to validate the downscaling models, a total of 14 samples at two depths (Table 2) were employed. The remote sensing and in situ data acquisition dates are shown in Table 2, with in situ data from 1 February 2016, corresponding to remote sensing data from 2 February 2016, and in situ data from 1 August 2016, corresponding to remote sensing data from 4 August 2016.

Table 2. Paired Dates of the Acquired Satellite Imagery and Chl-a Measurements of the Study Area.

MODIS	Sentinel-2A MSI	Chl-a Measurements
2016.2.2	2016.2.2	2016.2.1
2016.8.4	2016.8.4	2016.8.1

3.2. Preprocessing

The image preprocessing methods the same as (1) MODIS gap filling, (2) resampling, (3) land-water separation, and (4) subsetting were used to prepare all satellite images for further processes. Then, feature selection was employed to determine the most important band combinations for Chl-a prediction. First, the grid-fill method [53] was used to fill the missing values in the MODIS Chl-a images. This software has high computational efficiency and fills missing values in an iterative relaxation scheme using Poisson's equation. For validation, random removal of the original Chl-a pixel values was conducted to provide data to validate the grid-fill method. For this purpose, 5% of the original data were deleted from the outside boundary of the study region and preserved for cross-validation. Then, the corresponding values of the deleted data were estimated by utilizing the grid-fill method, and the difference between the original pixel values and the reconstructed values was evaluated. Consequently, the spatial distribution of Chl-a concentrations was produced.

At the second step, all S-2A images were mosaicked and then resampled to a resolution of 10 m (X_{10}). Since the study region had a remarkable number of islands (Figure 1), the reflectance caused by islands could decrease the accuracy of the applied models. Therefore, the third step of preprocessing aimed to separate water from land pixels to improve the downscaling results. The most commonly used indices, namely, the normalized difference vegetation index (NDVI) and normalized difference water index (NDWI), have already been employed for land-water separation purposes [54–58]. In the current study, land-water separation was performed by utilizing the NDWI index. The formula of this index is as follows:

$$\text{NDWI} = \frac{\rho_{\text{Green}} - \rho_{\text{NIR}}}{\rho_{\text{Green}} + \rho_{\text{NIR}}} \tag{1}$$

where ρ_{Green} and ρ_{NIR} are the green and near-infrared (NIR) reflectances of S-2A with 10 m resolution. Based on the NDWI computed images, water pixels have positive values, while negative pixels are usually classified as vegetation or soil features.

At the fourth step, the study area boundaries were used to produce a spatial subset of both MODIS Chl-a and S-2A images, and the S-2A dataset was resampled from 10 m (X_{10}) to 4 km resolution of MODIS Chl-a (X_{4k}) using the nearest neighbor method [59]. Then, MODIS Chl-a and S-2A images at 4 km (Y_{4k} and X_{4k}) were used as dependent and independent variables, respectively, to develop GP and MPR models. Additionally, the developed models were applied to S-2A images at 10 m (X_{10}) to estimate downscaled Chl-a at 10 m spatial resolution.

At the fifth step, the SVM-RFE feature selection method was employed to specify the most important bands of the S-2A dataset (X_{4k}). This method is a widely used technique for feature selection that has been utilized in a wide variety of remote sensing research studies [60–62]. To perform feature selection, multiple mathematical operations, including multiplication, addition, subtraction, rationing, averaging, and square transformation, were used to calculate 597 various combinations of the resampled S-2A bands at 4 km, and the computed combinations served as the input for the feature selection method. Then, SVM-RFE was trained on the calculated dataset using the MODIS Chl-a concentration (Y_{4k}) and S-2A band combinations (X_{4k}) as the predictand and predictor variables, respectively. As a result, relevant combinations for the downscaling procedure were chosen.

3.3. Downscaling

The downscaling model used in the current study was a regression-based approach considering the statistical relationship between MODIS Chl-a and S-2A bands at 4 km and 10 m spatial resolution. The straightforward formulation of this relationship is expressed as in Equation (2).

$$\hat{Y} = f(X) \qquad (2)$$

where X is the selected S-2A bands (predictor variables) at 4 km (X_{4k}) and 10 m (X_{10}), $\hat{Y}$ is the estimated MODIS Chl-a at 4 km ($\hat{Y}_{4k}$) and 10 m ($\hat{Y}_{10}$), and f represents a nonlinear regression function developed by GP and MPR techniques.

First, a relationship between Y_{4k} and X_{4k} was established using all the applied methods as 2nd-degree MPR, 3rd-degree MPR, 4th-degree MPR, and GP. From the 636 pixels of the independent (X_{4k}) and dependent (Y_{4k}) variables, 445 pixels were used for training, and the remaining 191 pixels were reserved for the validation of the models. Standardization was a prerequisite step before training all the applied methods to ensure that all variables remained on the same scale. Therefore, the standardization of all independent variables (X_{4k}) in the training and validation set was performed by subtracting the mean and dividing by the standard deviation of the training set. This preprocessing sped up the convergence and allowed efficient training of the network. Then, the estimated Chl-a at 4 km ($\hat{Y}_{4k}$) was subtracted from the original MODIS Chl-a at 4 km (Y_{4k}) as follows:

$$\varepsilon_{4k} = Y_{4k} - \hat{Y}_{4k} \qquad (3)$$

where ε_{4k} is the low-resolution residual at 4 km.

A simple kriging interpolation technique [63,64] was utilized to interpolate the ε_{4k} to 10 m resolution (ε_{10}) using the center points of the MODIS Chl-a pixels. Finally, to produce the downscaled map of Chl-a at 10 m resolution (Y_{10}), the developed model as $\hat{Y}_{10} = f(X_{10})$ in Equation (2) was added to ε_{10}, which is expressed in the function below (Equation (4)):

$$Y_{10} = \hat{Y}_{10} + \varepsilon_{10}. \qquad (4)$$

3.4. Sensitivity Analysis

For a given mathematical model, a sensitivity analysis (SA) measures how much the uncertainty and fluctuations of the input variable contribute to the outputs or performance of the system. In general, SA may be performed by two different techniques, local and global SA, with the former exploring the important model factors for a given set of factor values, and the latter apportioning the uncertainty in outputs to the uncertainty in each input factor to identify the important factors [65]. In the current study, the Morris method [66], as the most widely used global SA method, was employed to quantify the sensitivity of the GP and MPR models. This method is also called the once-at-a-time (OAT) method because, in each run, a new value is assigned to only one input variable. To carry out SA, the Morris sensitivity measure index (μ^*) was used as in Equation (5):

$$\mu_i^* = \frac{\sum_{n=1}^{r} |EE_{i,n}|}{r} \qquad (5)$$

where i is the number of input variables, r is the number of sample points in the parameter space (indexed n) and $EE_{i,n}$ is the elementary effects (EEs) assessed for the i-th input variable using the n-th sample point. EEs are employed to specify noninfluential inputs for a computationally costly mathematical model or for a model with a great number of input parameters, where the costs of evaluating other SA methods such as variance-based methods are not reasonably priced.

3.5. Mathematical Background

3.5.1. Multiple Polynomial Regression

The polynomial model is a form of a regression method that can be used when the relationship between an independent and dependent variable is curvilinear [67]. The nth order polynomial model with one variable Equation (6) is the general form of the polynomial model that indicates the nonlinear relationship between one predictand and one predictor variable.

$$Y = \beta_0 + \beta_1 X + \beta_2 X^2 + \ldots + \beta_m X^n + \varepsilon \tag{6}$$

where $\beta_1, \beta_2, \ldots, \beta_m$ are the unknown regression coefficients and ε is random error. Furthermore, MPR can also be defined with different degrees. For instance, a quadratic (second-order, n = 2) polynomial model can be given as in Equation (7).

$$Y = \beta_0 + \beta_1 X_1 + \beta_2 X_2 + \beta_{11} X_1^2 + \beta_{22} X_2^2 + \beta_{12} X_1 X_2 + \varepsilon \tag{7}$$

To select the best model between different MPR degrees for MODIS Chl-a downscaling, three degrees of models, including second-order (2nd), third-order (3rd), and fourth-order (4th), were trained and compared.

3.5.2. Genetic Programming

GP is one of the most recent data-driven techniques developed by Koza [68] and is a collection of techniques for finding a highly fit individual in the space of possible solutions. In GP, individuals are mathematical formulas created by combinations of functions (e.g., *sin*, *cos*, ÷, ×, +) and variables (e.g., *x*, *y*, 6). Each individual takes the role of a possible solution for a given problem. Figure 3 shows four simple formulas (individuals) created by functions and variables as an example. Each individual has its fitness value to optimize. GP applies evolutionary computation to find the best individual for optimized fitness values.

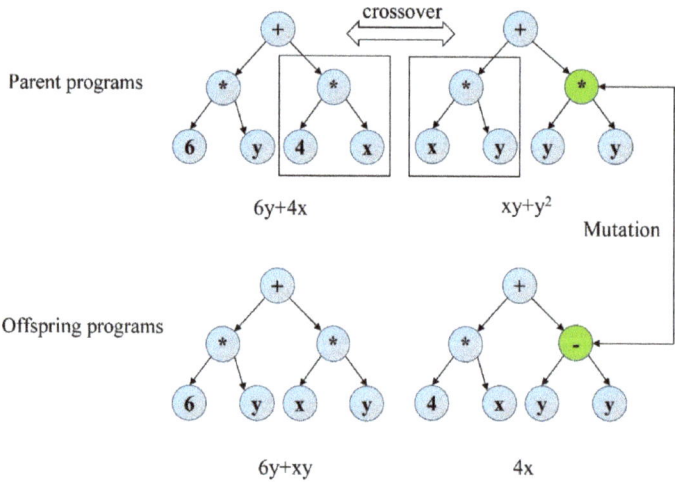

Figure 3. Tree structures illustrating crossover and mutation in GP.

Generally, the GP technique follows four steps to find the fittest individual: (1) an initial random population of individuals composed of functions and variables is created; (2) the fitness of each individual in the population is validated with a problem-specific fitness function, and the most appropriate individuals are selected to survive in the new population as parents; (3) once parents are selected, they create better types known as offspring or new generations by producing algorithms known as genetic operators (i.e., crossover, mutation, and duplication); (4) then, the individuals are assessed for fitness; (5) the process from (2) to (4) is repeated over many generations until an individual satisfies a given success criterion (e.g., the number of generations exceeds the specified number of iterations).

Figure 3 illustrates the crossover and mutation operations in GP. Individuals in GP are shown in tree form with different characteristics, such as size, shape, and content. The crossover and mutation operations are performed to produce new offspring (the panels in the lower row) from the parent (the panels in the upper row). Additionally, Figure 3 displays how crossover and mutation change the final functions and eliminate an input variable y at the second offspring (the lower right panel).

In the current study, the gplearn package [69] in Python software was used for GP implementation. In general, GP has two major parameters (population size and generation size) that should be optimized to generate high performance. The optimum values of the mentioned parameters were computed by utilizing the harmony search (HS) algorithm introduced by Geem, et al. [70] using 10-fold cross-validation. For more information about the HS method, readers are referred to Geem, Kim and Loganathan [70] and Lee and Singh [59].

4. Results

4.1. Preprocessing

The scatter plots between the original MODIS Chl-a and their reconstructed values with the grid-fill method are shown in Figure 4. According to the results, a good match between the original MODIS Chl-a and their filled values can be seen with $R^2 = 0.996$ for winter (panel (a)) and $R^2 = 0.939$ for summer (panel (b)). Therefore, it can be concluded that the grid-fill method exhibits good performance for the reconstruction of MODIS Chl-a values. Note that the reason for the better performance of the method on the winter day than on the summer day might be related to the presence of more missing values in the selected scene of the summer day that occurs due to high cloudiness on the summer day.

The most important combinations of S-2A images, among the 597 cases, were chosen by utilizing the SVR-RFE method. Four high-ranked combinations were selected as the final variables, namely, B1/B3, B2/B3, B1/(B3+B4), and B2/(B3+B4). These selected predictors are a combination of bands 1 (Coastal aerosol), 2 (Blue), 3 (Green), and 4 (Red).

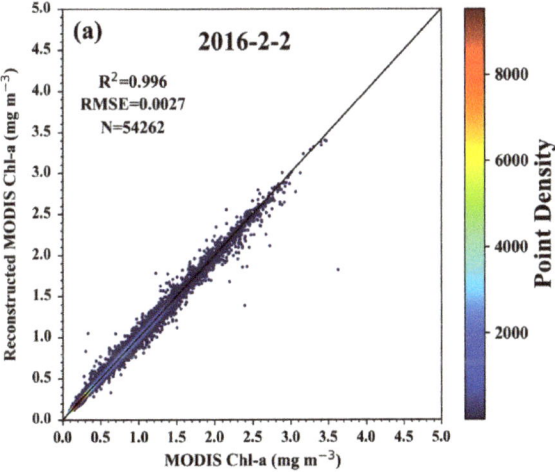

Figure 4. *Cont.*

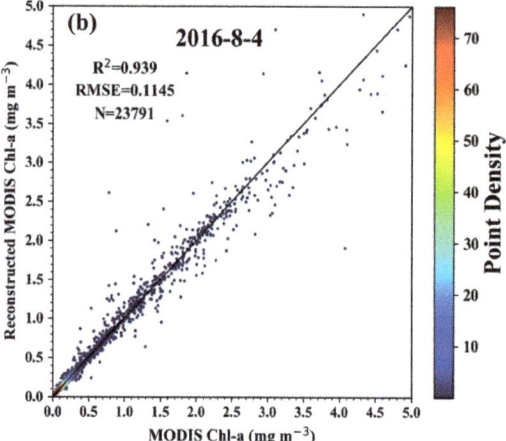

Figure 4. 1:1 scatter plots of the original MODIS Chl-a vs. reconstructed Chl-a by the grid-fill method on (**a**) the winter day (2016.2.2) and (**b**) the summer day (2016.8.4).

4.2. Downscaling Results

For downscaling, MPR with three degrees (2nd, 3rd, and 4th) and GP were trained with the four determined band combinations as predictors and MODIS Chl-a as a predictand at low-resolution (4 km). Then, a residual correction was performed utilizing the described methodology in the downscaling section (Section 3.3), and Equations (3)–(4) to produce high-resolution Chl-a maps (Y_{10}). To assess the accuracy of the downscaling technique, the final downscaled maps (i.e., Y_{10}) were validated with the original MODIS Chl-a maps at a pixel size of 4 km. For this purpose, sample values were extracted from the downscaled maps (Y_{10}) within a 3 × 3 window around the center of each MODIS Chl-a pixel, and the mean of each sample was calculated and compared with the original MODIS Chl-a (Figure 5 and Table 3).

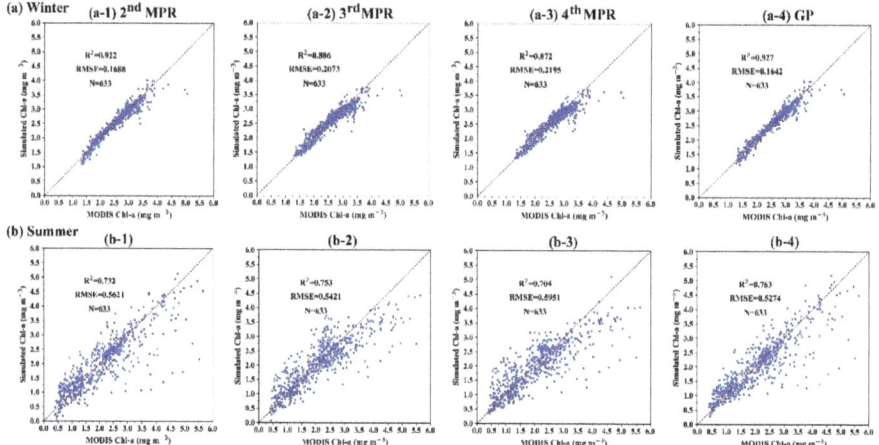

Figure 5. 1:1 scatter plots of the original MODIS Chl-a at 4 km pixel size vs. downscaled MODIS Chl-a at 10 m pixel size; (**a**) the winter day (2016.2.2): (**a-1**) 2nd-degree MPR, (**a-2**) 3rd-degree MPR, (**a-3**) 4th-degree MPR, (**a-4**) GP, and (**b**) the summer day (2016.8.4): (**b-1**) 2nd-degree MPR, (**b-2**) 3rd-degree MPR, (**b-3**) 4th-degree MPR, and (**b-4**) GP.

Table 3. Comparison of Performance Indices for All Models after Residual Correction.

Date	Performance Statistics	Model			
		2nd Degree MPR	3rd Degree MPR	4th Degree MPR	GP
2016.2.2 (Winter)	MAE	0.108	0.144	0.150	0.108
	MBE	−0.014	−0.012	−0.020	0.011
	RMSE	0.168	0.207	0.219	0.164
	R^2	0.922	0.886	0.872	0.927
2016.8.4 (Summer)	MAE	0.360	0.383	0.409	0.341
	MBE	−0.056	−0.044	−0.068	−0.030
	RMSE	0.562	0.542	0.595	0.527
	R^2	0.732	0.753	0.704	0.763

Notes: MAE is mean absolute error, MBE is mean bias error, RMSE is the root mean square error, and R^2 is the coefficient of determination.

On the winter day, as shown in Table 3, the performance measurements of the GP model were slightly better than those of the 2nd-degree MPR. According to the performance indices, the accuracy of the models was ranked as GP > 2nd-degree MPR > 3rd-degree MPR > 4th-degree MPR for the winter day. For the summer day, the rank was GP > 3rd-degree MPR > 2nd-degree MPR > 4th-degree MPR. Overall, the GP exhibited the best performance for the winter and summer days.

This finding shows the superiority of GP over the MPR method with different degrees of Chl-a downscaling. The possible reason for this result might be related to the flexible structure of the GP model compared with the fixed formulation of MPR models. While MPR models use a fixed form to define the relationship between the predictor and predictand variables, the evolution process in GP allows its function to take any feasible formulation. This flexibility gives GP the ability to adopt any form of functions to capture various relationships between the predictor and predictand variables, even highly nonlinear relations. Therefore, this unique feature of GP increases the probability of finding the best relationship in the downscaling procedure, resulting in a better prediction than the MPR models.

The accuracy of the models is presented in Figure 5. The best match between the MODIS Chl-a and simulated values can be seen in the GP model (see the panels in the fourth column in Figure 5), with $R^2 = 0.927$ and RMSE = 0.1642 on the winter day, compared to the MPR models (2nd-, 3rd-, and 4th-degree). The same result can be seen on the summer day (the best performance in the GP model), as shown in the bottom panels of Figure 5. Furthermore, it can be seen that all the applied models estimate Chl-a better at lower concentrations than at higher concentrations, particularly in the range of 1.5–3.5 mg m^{-3}, as presented in Figure 5.

4.3. Model Evaluation

4.3.1. Visual Comparison

The detailed maps of the downscaling approach for the winter and summer days are shown in Figures 6 and 7, respectively. The good agreement between the original MODIS Chl-a (the panels in the first column in Figure 6) and the estimated Chl-a maps (the panels in the last column in Figure 6) can be seen for the GP and 2nd-degree MPR. From Figure 6, the major trend of Chl-a is fairly modeled with GP and 2nd-degree MPR but specific and substantially high values are not captured in both models such as the values in the near coastal area. However, the residual model can additionally capture this high variability and produce a reliable estimate in the final stage, as seen in the last column of Figure 6. According to the results illustrated in Table 3, and the panels in the last column of Figure 6, 4th-degree MPR cannot estimate the Chl-a values at 10 m resolution as accurately as the other models, especially in coastal areas. This phenomenon indicates that the GP and 2nd-degree MPR can capture the most variation in high Chl-a concentrations at 4 km resolution.

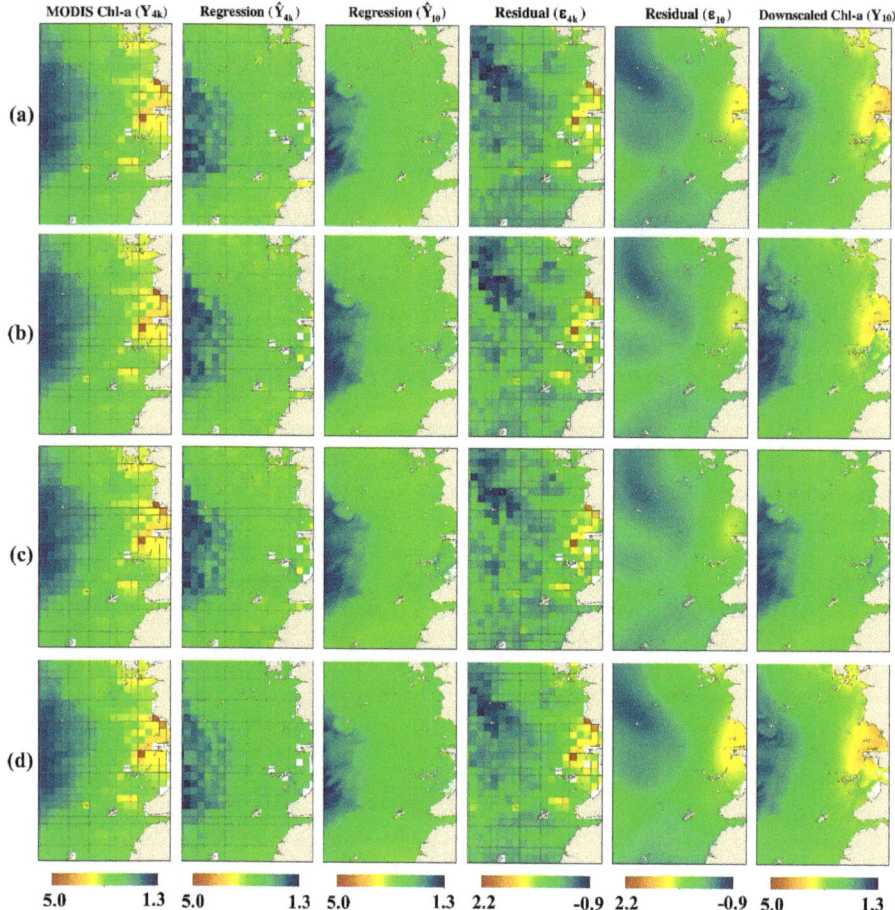

Figure 6. Detailed maps showing the downscaling steps for all models on the winter day (2016.2.2); (**a**) 2nd-degree MPR, (**b**) 3rd-degree MPR, (**c**) 4th-degree MPR, and (**d**) GP.

By estimating the Chl-a concentrations on the summer day, similar model behaviors became obvious on the summer day compared to the winter day (Figure 7). Among all the applied models, GP estimations in coastal areas showed better agreement with the original MODIS Chl-a on both the winter and summer days than all the applied MPR models. Since coastal areas play a vital role in marine ecosystems and human health, GP estimations are considerable for water quality monitoring in coastal areas.

From Figure 7, the simulation results became worse in the sea (the second and last column panels in Figure 7), and all models tended to underestimate some high Chl-a values. The possible reason for this result might be the higher spatial variability of the MODIS Chl-a concentration on the summer day than on the winter day (Figure 8), which is different from the normal distribution ($\delta = 0.522$, $\theta = 0.733$). Therefore, all the applied models did not fairly estimate the Chl-a values in the sea, and their accuracy decreased on the summer day compared to that on the winter day (Figure 5 and Table 3). From Figures 6 and 7, the Chl-a concentration gradient follows a similar pattern with water depth in all maps, and its concentration decreases as water depth increases. Therefore, the region close to the seashore shows a high Chl-a concentration, while the region in the sea presents a low concentration of Chl-a.

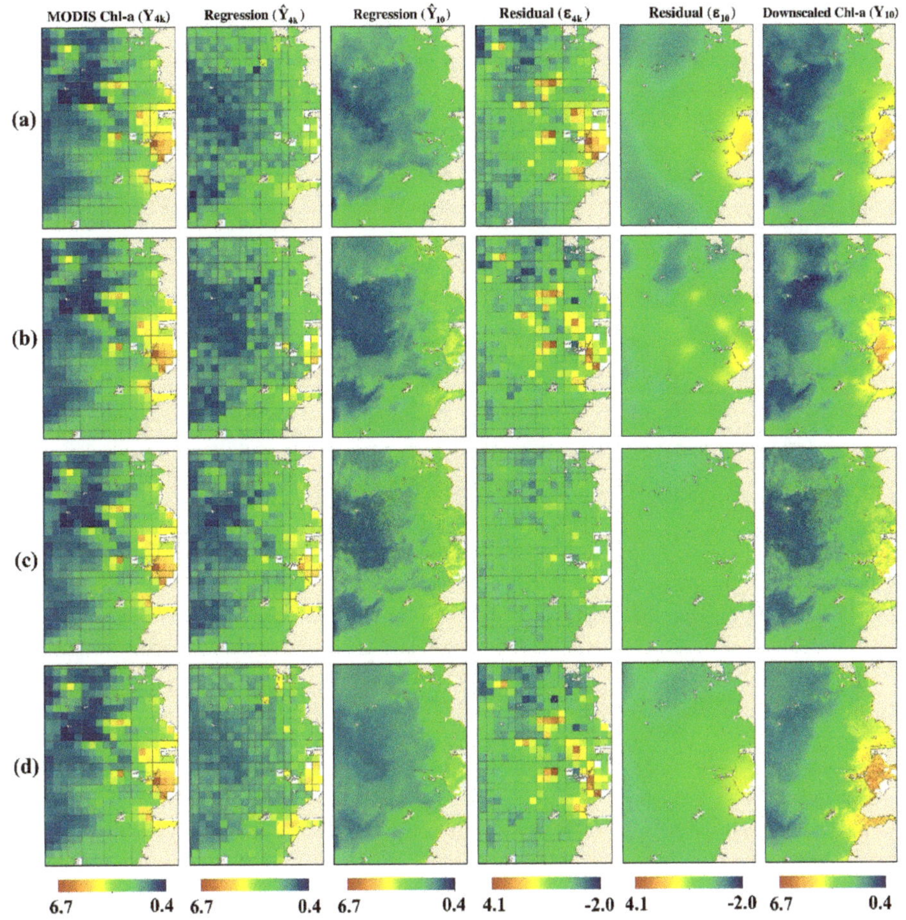

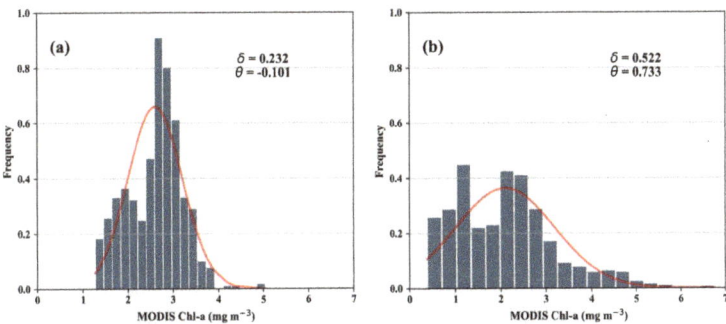

Figure 7. Detailed maps showing the downscaling steps for all models on the summer day (2016.8.4); (**a**) 2nd-degree MPR, (**b**) 3rd-degree MPR, (**c**) 4th-degree MPR, and (**d**) GP.

Figure 8. Histogram and fitted normal distribution of MODIS Chl-a for (**a**) the winter day (2016.2.2) and (**b**) the summer day (2016.8.4). δ and θ are the coefficient of variation (COV) and skewness coefficient, respectively. Note that the COV and skewness coefficient values for a normal distribution are zero. The skewness coefficient indicates that the distribution of the summer day data presents more non-normality than that of the winter day.

4.3.2. In Situ Validation

The results of the downscaling models were assessed with the in situ Chl-a data of seven stations along the study area (Figure 1) to validate model performances. For this purpose, sample values were extracted within a 3 × 3 window around the center of each station point, and the mean of each sample was calculated and compared with the in situ data. The results in terms of the R^2 and RMSE are presented in Figure 9 for the winter day (left panel) and the summer day (right panel). The computed p-values (Figure 9) shows that R^2 of all models is statistically significant (p-values lower than 0.05). In general, the R^2 of the GP model was higher than that of the other models for the winter and summer days, which were 0.59 and 0.47, respectively, as shown in Figure 9. Additionally, the RMSE of the GP model was smaller than that of the other models for the winter and summer days, at 0.766 and 0.483, respectively.

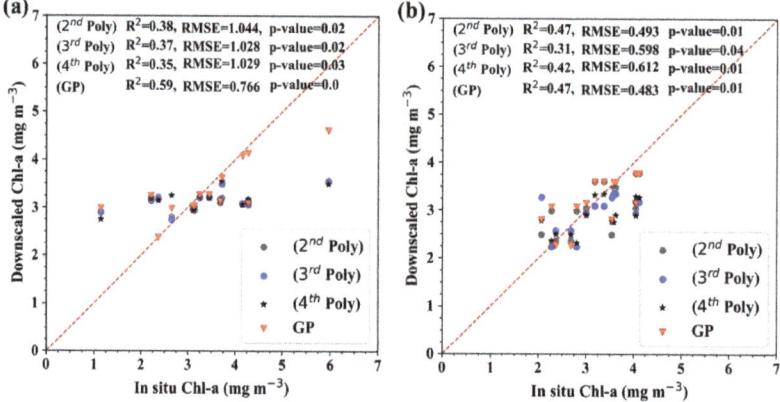

Figure 9. 1:1 scatter plots of downscaled MODIS Chl-a and in situ data for all methods using two measured data at 0–15 cm depth; (**a**) the winter day (2016.2.2) and (**b**) the summer day (2016.8.4).

These results indicate that the GP model can estimate Chl-a concentrations more accurately than the other models on both winter and summer days. From Figures 5 and 9, one can see that the validation of the downscaling models with the original MODIS Chl-a and the in situ data provides different levels of accuracy. This discrepancy shows the uncertainties in remote sensing data, indicating that the performance of the downscaling model is greatly affected by the accuracy of the remote sensing reflectance.

4.3.3. Sensitivity Analysis

To explore the sensitivity of the applied models to the changes in input parameters, the sensitivity measure index (μ^*) was calculated, as shown in Equation (5), for all predictor variables on both winter and summer days (Figures 10 and 11). The results of SA indicate that all the applied models are sensitive to the surface reflectance changes (μ^* values range from 0 to 1.45 for the winter and 0 to 2.42 for the summer day), and this sensitivity varies between the applied models. Figures 10 and 11 show that the 2nd-degree MPR model is the most sensitive model (μ^* values ranged from 0.09 to 1.24 for the winter and 0.94 to 2.42 for the summer day), while the GP model is the least sensitive model ($\mu^* = 1.45$ and 1.4 for B2/B3 band combination in winter and summer days, respectively) to the changes in the predictor variables. In addition, the sensitivity of the MPR models increases on the summer day compared to that of the winter day, while the GP model shows a slight decrease to the changes of the B2/B3 band combination on the summer day compared to that of the winter day.

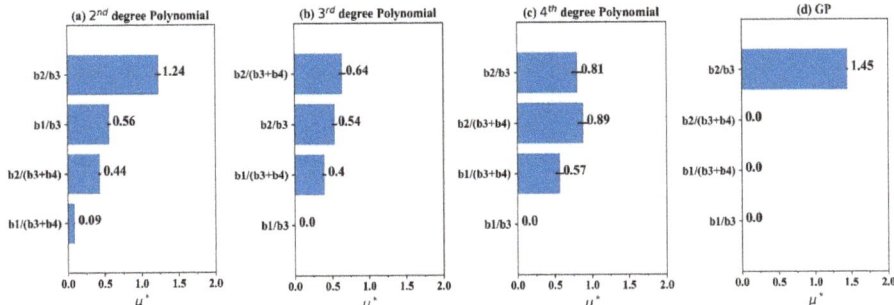

Figure 10. Sensitivity measure index (μ^*) for the winter day (2016.2.2); (**a**) 2nd-degree MPR, (**b**) 3rd-degree MPR, (**c**) 4th-degree MPR, and (**d**) GP. Note that the higher values of μ^* for a given parameter indicate higher sensitivity of the model to the changes of the parameter.

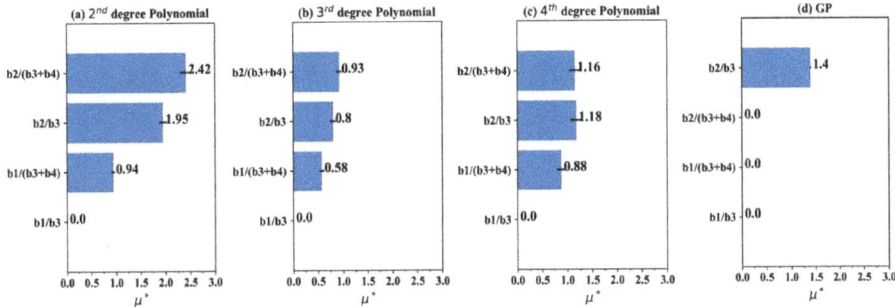

Figure 11. Sensitivity measure index (μ^*) for the summer day (2016.8.4); (**a**) 2nd-degree MPR, (**b**) 3rd-degree MPR, (**c**) 4th-degree MPR, and (**d**) GP. Note that the higher values of μ^* for a given parameter indicate higher sensitivity of the model to the changes of the parameter.

It can be concluded that although GP and 2nd-degree MPR show comparable accuracy (Figure 5 and Table 3) on both winter and summer days, the GP model with the least sensitivity to the changes in the input parameters is more effective than the 2nd-degree MPR for downscaling Chl-a. Additionally, SA results show that the accuracy of all models is strongly related to the accuracy of the remote sensing data; this further confirms the need for atmospheric correction as an essential task in the downscaling procedure [71].

5. Discussion and Conclusions

In the current study, a downscaling framework composed of GP and MPR models was developed to downscale MODIS Chl-a from 4 km to 10 m pixel size using S-2A band combinations at 10 m as predictor variables. The MODIS Chl-a at 4 km was downscaled in four main steps: (i) acquiring MODIS Chl-a and S-2A images at 4 km and 10 m, respectively; (ii) applying feature selection to select the most important S-2A band combinations as predictor variables; (iii) employing the trained MPR with three degrees (2nd, 3rd, and 4th) and GP model to downscale MODIS Chl-a; (iv) assessment of the results with original MODIS Chl-a maps at a pixel size of 4 km and in situ measurements.

By comparing the performance of all the models applied for downscaling, the GP model presents more accurate results and less sensitivity to the changes in all variables for downscaling Chl-a. The superiority of the GP model over MPR models is related to the flexible formulation structure of the GP compared to the fixed formulation of the MPR models. The flexible structure of the GP allows its function to take any feasible formulation so that it increases the probability of finding the best input-output relationship between thousands of formula. Visual comparison of the applied models showed that although the major trend of Chl-a is fairly modeled with GP and 2nd-degree MPR, but specific and substantially high values are not captured with both models such as the values in near coastal areas. This drawback of the models can be solved with the residual correction that makes it an essential procedure to improve the accuracy of the spatial downscaling model.

Among all the applied models, the GP model provided better estimations in coastal areas than all the MPR models. Therefore, GP can serve as a feasible alternative model to estimate Chl-a concentrations in coastal areas with complex characteristics, where water quality monitoring plays a vital role in the protection of marine ecosystems and human health. Moreover, the results indicate that the performance of the models greatly depends on the spatial variability of the MODIS Chl-a concentration. Its distributional characteristics (e.g., normal or skewed) can be a good option for model selection criteria. Therefore, to ensure the validity of the GP model in other coastal areas, it is recommended to assess the normality to select the GP model for spatial downscaling of the MODIS Chl-a.

Although the downscaling procedure used in the current study was applied on two selected days, one in winter and one in summer based on data availability, it produced reasonable results due to using the S-2A dataset with 10 m spatial resolution. Future studies can be focused on extending this procedure at a higher temporal resolution. In addition, more research can be conducted with deep learning techniques for downscaling Chl-a concertation. In addition to the surface reflectance, more predictors such as NDVI and NDWI, due to the strong Chl-a absorption at red-NIR spectral regions, may be used in determining Chl-a concentration.

Author Contributions: H.M. initiated the idea and contributed to the analysis of the data and writing of the manuscript. T.L. secured funding for the project and contributed to the writing of the manuscript. J.Y. made revisions to the manuscript. All authors have read and agreed to the published version of the manuscript.

Funding: This research was funded by National Research Foundation of Korea (NRF), grant number 2018R1A2B6001799.

Acknowledgments: The authors would like to thank NASA and the ESA (European Space Agency) for providing the MODIS and Sentinel-2A MSI data, respectively. Additionally, we appreciate the Korea Institute of Ocean Science & Technology (KIOST) for providing Chl-a measurements of the western coast of South Korea. We would like to thank the three anonymous reviewers, whose comments significantly improved the paper.

Conflicts of Interest: The authors declare no conflict of interest.

References

1. Bierman, P.; Lewis, M.; Ostendorf, B.; Tanner, J. A review of methods for analysing spatial and temporal patterns in coastal water quality. *Ecol. Indic.* **2011**, *11*, 103–114. [CrossRef]
2. Blondeau-Patissier, D.; Gower, J.F.; Dekker, A.; Phinn, S.; Brando, V. A review of ocean color remote sensing methods and statistical techniques for the detection, mapping and analysis of phytoplankton blooms in coastal and open oceans. *Prog. Oceanogr.* **2014**, *123*, 123–144. [CrossRef]
3. Cullen, J.J. The deep chlorophyll maximum: Comparing vertical profiles of chlorophyll a. *Can. J. Fish. Aquat. Sci.* **1982**, *39*, 791–803. [CrossRef]
4. Guan, X.; Li, J.; Booty, W.G. Monitoring lake simcoe water clarity using Landsat-5 TM images. *Water Resour. Manag.* **2011**, *25*, 2015–2033. [CrossRef]
5. Carpenter, S.R.; Caraco, N.F.; Correll, D.L.; Howarth, R.W.; Sharpley, A.N.; Smith, V.H. Nonpoint pollution of surface waters with phosphorus and nitrogen. *Ecol. Appl.* **1998**, *8*, 559–568. [CrossRef]
6. Smith, V.H. Cultural eutrophication of inland, estuarine, and coastal waters. In *Successes, Limitations, and Frontiers in Ecosystem Science*; Springer: New York, NY, USA, 1998; pp. 7–49.
7. Kar, D. *Epizootic Ulcerative Fish Disease Syndrome*; Academic Press: Cambridge, MA, USA, 2015.
8. Bacher, C.; Grant, J.; Hawkins, A.J.; Fang, J.; Zhu, M.; Besnard, M. Modelling the effect of food depletion on scallop growth in Sungo Bay (China). *Aquat. Living Resour.* **2003**, *16*, 10–24. [CrossRef]
9. Huot, Y.; Babín, M.; Bruyant, F.; Grob, C.; Twardowski, M.S.; Claustre, H. Does chlorophyll a provide the best index of phytoplankton biomass for primary productivity studies? *Biogeosci. Discuss.* **2007**, *4*, 707–745. [CrossRef]

10. Goetz, S.J.; Gardiner, E.P.; Viers, J.H. Monitoring freshwater, estuarine and near-shore benthic ecosystems with multi-sensor remote sensing: An introduction to the special issue. *Remote Sens. Environ.* **2008**, *112*, 3993–3995. [CrossRef]
11. Kutser, T. Passive optical remote sensing of cyanobacteria and other intense phytoplankton blooms in coastal and inland waters. *Int. J. Remote Sens.* **2009**, *30*, 4401–4425. [CrossRef]
12. Schaeffer, B.A.; Schaeffer, K.G.; Keith, D.; Lunetta, R.S.; Conmy, R.; Gould, R.W. Barriers to adopting satellite remote sensing for water quality management. *Int. J. Remote Sens.* **2013**, *34*, 7534–7544. [CrossRef]
13. Merico, A.; Brown, C.W.; Groom, S.B.; Miller, P.; Tyrrell, T. Analysis of satellite imagery forEmiliania huxleyiblooms in the Bering Sea before 1997. *Geophys. Res. Lett.* **2003**, *30*, 30. [CrossRef]
14. Gower, J. Productivity and plankton blooms observed with Seawifs and in-situ sensors. In Proceedings of the IGARSS 2001, Scanning the Present and Resolving the Future, Proceedings, IEEE 2001 International Geoscience and Remote Sensing Symposium (Cat. No.01CH37217), Sydney, NSW, Australia, 9–13 July 2001; pp. 2181–2183.
15. Lavender, S.; Groom, S.B. The detection and mapping of algal blooms from space. *Int. J. Remote Sens.* **2001**, *22*, 197–201. [CrossRef]
16. Siegel, D.; Behrenfeld, M.J.; Maritorena, S.; McClain, C.; Antoine, D.; Bailey, S.W.; Bontempi, P.; Boss, E.; Dierssen, H.; Doney, S.C.; et al. Regional to global assessments of phytoplankton dynamics from the SeaWiFS mission. *Remote Sens. Environ.* **2013**, *135*, 77–91. [CrossRef]
17. Hu, C.; Muller-Karger, F.E.; Taylor, C.J.; Carder, K.L.; Kelble, C.; Johns, E.; Heil, C.A. Red tide detection and tracing using MODIS fluorescence data: A regional example in SW Florida coastal waters. *Remote Sens. Environ.* **2005**, *97*, 311–321. [CrossRef]
18. Gower, J.; King, S.; Yan, W.; Borstad, G.; Brown, L. Use of the 709 nm band of MERIS to detect intense plankton blooms and other conditions in coastal waters. In Proceedings of the MERIS User Workshop, Frascati, Italy, 10–13 November 2003.
19. Hu, C. A novel ocean color index to detect floating algae in the global oceans. *Remote Sens. Environ.* **2009**, *113*, 2118–2129. [CrossRef]
20. Claustre, H.; Babin, M.; Merien, D.; Ras, J.; Prieur, L.; Dallot, S.; Prášil, O.; Dousova, H.; Moutin, T. Toward a taxon-specific parameterization of bio-optical models of primary production: A case study in the North Atlantic. *J. Geophys. Res. Space Phys.* **2005**, *110*, 110. [CrossRef]
21. Sathyendranath, S.; Watts, L.; Devred, E.; Platt, T.; Caverhill, C.; Maass, H. Discrimination of diatoms from other phytoplankton using ocean-colour data. *Mar. Ecol. Prog. Ser.* **2004**, *272*, 59–68. [CrossRef]
22. Shang, S.; Dong, Q.; Lee, Z.; Li, Y.; Xie, Y.; Behrenfeld, M. MODIS observed phytoplankton dynamics in the Taiwan Strait: An absorption-based analysis. *Biogeosciences* **2011**, *8*, 841–850. [CrossRef]
23. Lee, Z.; Carder, K.L.; Arnone, R.A. Deriving inherent optical properties from water color: A multiband quasi-analytical algorithm for optically deep waters. *Appl. Opt.* **2002**, *41*, 5755–5772. [CrossRef]
24. Addesso, P.; Longo, M.; Maltese, A.; Restaino, R.; Vivone, G. Batch methods for resolution enhancement of TIR image sequences. *IEEE J. Sel. Top. Appl. Earth Obs. Remote Sens.* **2015**, *8*, 3372–3385. [CrossRef]
25. Bechtel, B.; Zakšek, K.; Hoshyaripour, G. Downscaling land surface temperature in an urban area: A case study for Hamburg, Germany. *Remote Sens.* **2012**, *4*, 3184–3200. [CrossRef]
26. Liu, D.; Pu, R. Downscaling thermal infrared radiance for subpixel land surface temperature retrieval. *Sensors* **2008**, *8*, 2695–2706. [CrossRef] [PubMed]
27. Chen, C.; Zhao, S.; Duan, Z.; Qin, Z. An improved spatial downscaling procedure for TRMM 3B43 precipitation product using geographically weighted regression. *IEEE J. Sel. Top. Appl. Earth Obs. Remote Sens.* **2015**, *8*, 4592–4604. [CrossRef]
28. Fang, J.; Du, J.; Xu, W.; Shi, P.; Li, M.; Ming, X. Spatial downscaling of TRMM precipitation data based on the orographical effect and meteorological conditions in a mountainous area. *Adv. Water Resour.* **2013**, *61*, 42–50. [CrossRef]
29. Immerzeel, W.; Rutten, M.; Droogers, P. Spatial downscaling of TRMM precipitation using vegetative response on the Iberian Peninsula. *Remote Sens. Environ.* **2009**, *113*, 362–370. [CrossRef]
30. Zhang, T.; Li, B.; Yuan, Y.; Gao, X.; Sun, Q.; Xu, L.; Jiang, Y. Spatial downscaling of TRMM precipitation data considering the impacts of macro-geographical factors and local elevation in the Three-River Headwaters Region. *Remote Sens. Environ.* **2018**, *215*, 109–127. [CrossRef]

31. Kaheil, Y.H.; Gill, M.K.; Bastidas, L.A.; Rosero, E.; McKee, M. Downscaling and assimilation of surface soil moisture using ground truth measurements. *IEEE Trans. Geosci. Remote Sens.* **2008**, *46*, 1375–1384. [CrossRef]
32. Shi, J.; Jiang, L.; Zhang, L.; Chen, K.; Wigneron, J.-P.; Chanzy, A.; Jackson, T. Physically based estimation of bare-surface soil moisture with the passive radiometers. *IEEE Trans. Geosci. Remote Sens.* **2006**, *44*, 3145–3153. [CrossRef]
33. Fu, Y.; Xu, S.; Zhang, C.; Sun, Y. Spatial downscaling of MODIS Chlorophyll-a using Landsat 8 images for complex coastal water monitoring. *Estuar. Coast. Shelf Sci.* **2018**, *209*, 149–159. [CrossRef]
34. Gao, F.; Kustas, W.; Anderson, M.C. A data mining approach for sharpening thermal satellite imagery over land. *Remote Sens.* **2012**, *4*, 3287–3319. [CrossRef]
35. Ghosh, A.; Joshi, P. Hyperspectral imagery for disaggregation of land surface temperature with selected regression algorithms over different land use land cover scenes. *ISPRS J. Photogramm. Remote Sens.* **2014**, *96*, 76–93. [CrossRef]
36. Hutengs, C.; Vohland, M. Downscaling land surface temperatures at regional scales with random forest regression. *Remote Sens. Environ.* **2016**, *178*, 127–141. [CrossRef]
37. Mehr, A.D.; Nourani, V.; Kahya, E.; Hrnjica, B.; Sattar, A.M.A.; Yaseen, Z.M. Genetic programming in water resources engineering: A state-of-the-art review. *J. Hydrol.* **2018**, *566*, 643–667. [CrossRef]
38. Koh, C.-H.; Khim, J.S. The Korean tidal flat of the Yellow Sea: Physical setting, ecosystem and management. *Ocean Coast. Manag.* **2014**, *102*, 398–414. [CrossRef]
39. Park, K.-A.; Lee, E.-Y.; Chang, E.; Hong, S. Spatial and temporal variability of sea surface temperature and warming trends in the Yellow Sea. *J. Mar. Syst.* **2015**, *143*, 24–38. [CrossRef]
40. Zhang, C.I.; Lim, J.-H.; Kwon, Y.; Kang, H.J.; Kim, H.; Seo, Y.I. The current status of west sea fisheries resources and utilization in the context of fishery management of Korea. *Ocean Coast. Manag.* **2014**, *102*, 493–505. [CrossRef]
41. Zhang, C.; Kim, S. Living marine resources of the Yellow Sea ecosystem in Korean waters: Status and perspectives. In *Large Marine Ecosystems of the Pacific Rim*; Wiley, Blackwell Science: Cambridge, MA, USA, 1999; pp. 163–178.
42. Ye, H.; Li, J.; Li, T.; Shen, Q.; Zhu, J.; Wang, X.; Zhang, F.; Zhang, J.; Zhang, B. Spectral classification of the Yellow Sea and implications for coastal ocean color remote sensing. *Remote Sens.* **2016**, *8*, 321. [CrossRef]
43. Cho, Y.-K.; Kim, M.-O.; Kim, B.-C. Sea fog around the Korean Peninsula. *J. Appl. Meteorol.* **2000**, *39*, 2473–2479. [CrossRef]
44. Letelier, R.M. An analysis of chlorophyll fluorescence algorithms for the moderate resolution imaging spectrometer (MODIS). *Remote Sens. Environ.* **1996**, *58*, 215–223. [CrossRef]
45. Sarthyendranath, S. *Remote Sensing of Ocean Colour in Coastal, and Other Optically-Complex, Waters*; International Ocean Colour Coordinating Group (IOCCG): Dartmouth, NS, Canada, 2000.
46. Hu, C.; Lee, Z.; Franz, B. Chlorophyll a algorithms for oligotrophic oceans: A novel approach based on three-band reflectance difference. *J. Geophys. Res. Ocean.* **2012**, *117*. [CrossRef]
47. Pahlevan, N.; Schott, J.R. Leveraging EO-1 to evaluate capability of new generation of landsat sensors for coastal/inland water studies. *IEEE J. Sel. Top. Appl. Earth Obs. Remote Sens.* **2013**, *6*, 360–374. [CrossRef]
48. Pahlevan, N.; Lee, Z.; Wei, J.; Schaaf, C.; Schott, J.R.; Berk, A. On-orbit radiometric characterization of OLI (Landsat-8) for applications in aquatic remote sensing. *Remote Sens. Environ.* **2014**, *154*, 272–284. [CrossRef]
49. Gower, J.; King, S.; Borstad, G.; Brown, L. Detection of intense plankton blooms using the 709 nm band of the MERIS imaging spectrometer. *Int. J. Remote Sens.* **2005**, *26*, 2005–2012. [CrossRef]
50. Gower, J.; King, S.; Borstad, G.; Brown, L. The importance of a band at 709 nm for interpreting water-leaving spectral radiance. *Can. J. Remote Sens.* **2008**, *34*, 287–295.
51. Moses, W.J.; Gitelson, A.; Berdnikov, S.; Povazhnyy, V. Satellite estimation of chlorophyll-a concentration using the red and NIR bands of MERIS—The Azov Sea case study. *IEEE Geosci. Remote Sens. Lett.* **2009**, *6*, 845–849. [CrossRef]
52. Pahlevan, N.; Chittimalli, S.K.; Balasubramanian, S.V.; Vellucci, V. Sentinel-2/Landsat-8 product consistency and implications for monitoring aquatic systems. *Remote Sens. Environ.* **2019**, *220*, 19–29. [CrossRef]
53. Dawson, A. (2018, April 23). ajdawson/gridfill: Version 1.0.1 (Version v1.0.1). Zenodo.
54. Gao, H.; Birkett, C.; Lettenmaier, D.P. Global monitoring of large reservoir storage from satellite remote sensing. *Water Resour. Res.* **2012**, *48*, 48. [CrossRef]

55. McFeeters, S.K. The use of the Normalized Difference Water Index (NDWI) in the delineation of open water features. *Int. J. Remote Sens.* **1996**, *17*, 1425–1432. [CrossRef]
56. Mohebzadeh, H.; Fallah, M. Quantitative analysis of water balance components in Lake Urmia, Iran using remote sensing technology. *Remote Sens. Appl. Soc. Environ.* **2019**, *13*, 389–400. [CrossRef]
57. Sima, S.; Ahmadalipour, A.; Tajrishy, M. Mapping surface temperature in a hyper-saline lake and investigating the effect of temperature distribution on the lake evaporation. *Remote Sens. Environ.* **2013**, *136*, 374–385. [CrossRef]
58. Mohebzadeh, H. Extracting A-L relationship for Urmia Lake, Iran using MODIS NDVI/NDWI indices. *J. Hydrogeol. Hydrol. Eng.* **2018**, *7*, 1. [CrossRef]
59. Lee, T.; Singh, V.P. *Statistical Downscaling for Hydrological and Environmental Applications*; CRC Press: Boca Raton, FL, USA, 2018; Volume 1, p. 165.
60. Tuia, D.; Pacifici, F.; Kanevski, M.; Emery, W.J. Classification of very high spatial resolution imagery using mathematical morphology and support vector machines. *IEEE Trans. Geosci. Remote Sens.* **2009**, *47*, 3866–3879. [CrossRef]
61. Zheng, Z.; Zeng, Y.; Li, S.; Huang, W. A new burn severity index based on land surface temperature and enhanced vegetation index. *Int. J. Appl. Earth Obs. Geoinf.* **2016**, *45*, 84–94. [CrossRef]
62. Ebrahimy, H.; Azadbakht, M. Downscaling MODIS land surface temperature over a heterogeneous area: An investigation of machine learning techniques, feature selection, and impacts of mixed pixels. *Comput. Geosci.* **2019**, *124*, 93–102. [CrossRef]
63. Rossi, R.E.; Dungan, J.L.; Beck, L.R. Kriging in the shadows: Geostatistical interpolation for remote sensing. *Remote Sens. Environ.* **1994**, *49*, 32–40. [CrossRef]
64. Karl, J. Spatial predictions of cover attributes of rangeland ecosystems using regression kriging and remote sensing. *Rangel. Ecol. Manag.* **2010**, *63*, 335–349. [CrossRef]
65. Petropoulos, G.; Srivastava, P.K. *Sensitivity Analysis in Earth Observation Modelling*; Elsevier: Amsterdam, The Netherlands, 2016.
66. Morris, M.D. Factorial sampling plans for preliminary computational experiments. *Technometrics* **1991**, *33*, 161–174. [CrossRef]
67. Rao, C.R.; Toutenburg, H.; Shalabh, H.C.; Schomaker, M. Linear models and generalizations. In *Least Squares and Alternatives*, 3rd ed.; Springer: Berlin/Heidelberg, Germany; New York, NY, USA, 2008.
68. Koza, J.R. *Genetic Programming: On the Programming of Computers by Means of Natural Selection*; MIT Press: Cambridge, MA, USA; London, UK, 1992; Volume 1.
69. Stephens, T. *Gplearn Model, Genetic Programming*; Copyright, 2015.
70. Geem, Z.W.; Kim, J.H.; Loganathan, G. A new heuristic optimization algorithm: Harmony search. *Simulation* **2001**, *76*, 60–68. [CrossRef]
71. Nazeer, M.; Nichol, J. Development and application of a remote sensing-based Chlorophyll-a concentration prediction model for complex coastal waters of Hong Kong. *J. Hydrol.* **2016**, *532*, 80–89. [CrossRef]

© 2020 by the authors. Licensee MDPI, Basel, Switzerland. This article is an open access article distributed under the terms and conditions of the Creative Commons Attribution (CC BY) license (http://creativecommons.org/licenses/by/4.0/).

Article

Capacity of the PERSIANN-CDR Product in Detecting Extreme Precipitation over Huai River Basin, China

Shanlei Sun [1,*], Jiazhi Wang [1], Wanrong Shi [1], Rongfan Chai [1] and Guojie Wang [2]

1 Collaborative Innovation Center on Forecast and Evaluation of Meteorological Disasters/Key Laboratory of Meteorological Disaster, Ministry of Education/International Joint Research Laboratory on Climate and Environment Change, Nanjing University of Information Science and Technology, Nanjing 210044, China; 20191201122@nuist.edu.cn (J.W.); 20181201117@nuist.edu.cn (W.S.); rfchai@nuist.edu.cn (R.C.)
2 School of Geographical Sciences, Nanjing University of Information Science and Technology, Nanjing 210044, China; gwang@nuist.edu.cn
* Correspondence: sun.s@nuist.edu.cn; Tel.: +86-025-5869-5622

Abstract: Assessing satellite-based precipitation product capacity for detecting precipitation and linear trends is fundamental for accurately knowing precipitation characteristics and changes, especially for regions with scarce and even no observations. In this study, we used daily gauge observations across the Huai River Basin (HRB) during 1983–2012 and four validation metrics to evaluate the Precipitation Estimation from Remotely Sensed Information Using Artificial Neural Networks-Climate Data Record (PERSIANN-CDR) capacity for detecting extreme precipitation and linear trends. The PERSIANN-CDR well captured climatologic characteristics of the precipitation amount- (PRCPTOT, R85p, R95p, and R99p), duration- (CDD and CWD), and frequency-based indices (R10mm, R20mm, and Rnnmm), followed by moderate performance for the intensity-based indices (Rx1day, R5xday, and SDII). Based on different validation metrics, the PERSIANN-CDR capacity to detect extreme precipitation varied spatially, and meanwhile the validation metric-based performance differed among these indices. Furthermore, evaluation of the PERSIANN-CDR linear trends indicated that this product had a much limited and even no capacity to represent extreme precipitation changes across the HRB. Briefly, this study provides a significant reference for PERSIANN-CDR developers to use to improve product accuracy from the perspective of extreme precipitation, and for potential users in the HRB.

Keywords: extreme precipitation index; PERSIANN-CDR; KGE; linear trend; Huai River Basin

Citation: Sun, S.; Wang, J.; Shi, W.; Chai, R.; Wang, G. Capacity of the PERSIANN-CDR Product in Detecting Extreme Precipitation over Huai River Basin, China. *Remote Sens.* 2021, 13, 1747. https://doi.org/10.3390/rs13091747

Academic Editor: Joo-Heon Lee

Received: 23 March 2021
Accepted: 28 April 2021
Published: 30 April 2021

Publisher's Note: MDPI stays neutral with regard to jurisdictional claims in published maps and institutional affiliations.

Copyright: © 2021 by the authors. Licensee MDPI, Basel, Switzerland. This article is an open access article distributed under the terms and conditions of the Creative Commons Attribution (CC BY) license (https://creativecommons.org/licenses/by/4.0/).

1. Introduction

With successive and rapid warming during the past decades, increasing evidence suggests that climate extremes (e.g., extreme precipitation, heatwaves, and droughts) have changed across the world [1]. Of climate extremes, extreme precipitation is believed to be one major cause of the water-related disasters, e.g., floods and landslides [2–5]. These water-related disasters often result in enormous loss of life and destruction and have become a major obstacle to the sustainable development of society and the economy [6–8]. The Global Emergency Disaster Database stated that from 1970 to 2013 across the world, more than ten thousand water-related disasters happened, impacting more than 6.6 billion people, and leading to more than USD 2600 billion in damage, with the death of 3.5 million people [9]. In one word, the adverse impact induced by extreme precipitation on life and socio-economy are enormous, and therefore it is very necessary and critical to understand extreme precipitation (e.g., spatial patterns, changes, and underlying mechanisms) to reduce the related disasters and to develop reasonable prevention strategies.

Despite that, studying extreme precipitation still presents immense challenges because of difficulties in obtaining accurate, uninterrupted, and uniform precipitation data at the regional and global scale. So far, three pathways are employed to measure precipitation, i.e., direct observation with various gauges and retrievals from radar and satellite techniques. Gauge precipitation is believed to be the most accurate measurement [10]. However, due to inaccessibility and higher costs for installations and maintenance, there are issues to using gauge precipitation, i.e., limited spatial representativeness and coverage, time discontinuity, and short time span [11]. In view of this, gauge precipitation cannot fully satisfy the specific requirements of academic studies and practical applications, e.g., long-term (>30 years), continuous (space and time) precipitation observations for climate studies. Radar-based precipitation retrievals have filled gaps in gauge precipitation to some extent, e.g., more extensive coverage. However, radar-based precipitation still has potential issues, e.g., the backwardness of radar technology in some countries and radar blockage due to topography [12–14]. In the past decades, very great advances have been made in technologies of satellites and sensors. Subsequently, various satellite-based precipitation products have been proposed based on radiance information received by satellite-carried sensors and different statistical and/or physics-based retrieval algorithms, such as Tropical Rainfall Measuring Mission (TRMM) Multi-Satellite Precipitation Analysis (GSFC; [15]); the National Oceanic and Atmospheric Administration (NOAA) Climate Prediction Center (CPC) morphing technique (CMORPH; [16]); the Global Satellite Mapping of Precipitation Microwave-IR Combined Product (GSMaP; [17]), the Precipitation Estimation from Remotely Sensed Information Using Artificial Neural Networks (PERSIANN; [18]); the Integrated Multisatellite Retrievals for Global Precipitation Measurement (IMERG; [19]); and the Climate Hazards Group InfraRed Precipitation with Station Data (CHIRPS; [20]). These satellite-based products provide an opportunity for academic studies and practical applications to fulfill various requirements of precipitation data.

Considering specific needs and goals, it is necessary to conduct quantitative evaluations of a given satellite-based precipitation product using dependable reference data, which can improve the confidence level of the related academic studies and ensure high efficiencies in practical applications. Studies have extensively assessed various satellite-based precipitation data over the world with a variety of statistical metrics [11,21–23]. For example, Shen et al. [21] suggested that CMORPH performed better than TRMMM and PERSIANN in capturing spatial and temporal variations in most of China, especially for reproducing summer precipitation characteristics. Tan et al. [22] compared multiple satellite-based precipitation estimates over Malaysia and found that TRMM showed higher coincidence with the observational precipitation. Results from Alijanian et al. [11] showed that Multi-Source Weighted-Ensemble Precipitation (MSWEP), PERSIANN-Climate Data Record (CDR), and TRMM could better identify rainfall and non-rainfall events in Iran, and PERSIANN-CDR had higher capacity than the other datasets in representing heavy rainfall. These studies provided references for theoretical understanding, and development of satellite-retrieved methodologies and their practical applications.

In China, more and more evidence indicates that extreme precipitation and related disasters have varied [24]. Taking the Huai River Basin (HRB) as an example, it frequently suffers from floods, with more than 350 floods during the past 50 decades and local and regional floods occurring nearly every two years [25]. Since the beginning of the 21st century, several severe extreme precipitation events (e.g., 2003, 2005, and 2007) and related floods happened in the HRB [25,26]. Incomplete statistics reported that more than 58 million people in the HRB were affected by the 2003 floods, with the flood-affected arable area exceeding 52,000 km^2, 390 thousand houses collapsed, and direct economic losses of more than CNY 35 billion [27]. Yin et al. [28] projected that extreme precipitation with return periods of 20- and 50-years during 2001–2050 would considerably increase with exacerbating global warming, particularly in some places with increases of more than 30%; this implies that the HRB has the potential to face a larger flood risk in future. Therefore, reasonably managing extreme precipitation-induced floods and taking efficient prevention measures are very important for regional and national food security and food production capacity. To this end, selecting a reliable, long-term, continuous precipitation measurement strategy with a relatively high spatio-temporal resolution is of much significance when attempting to mitigate the extreme precipitation-induced flood risk in the HRB; consequently, we chose the PERSIANN-CDR precipitation product here for evaluation with a high density of gauge records. Despite the fact that this product has been assessed over different regions of the world and even in China [23,29–32], issues still exist. For example, how good is the overall performance (e.g., Kling–Gupta Efficiency (KGE), which integrates impacts of bias, variability, and correlation coefficient on the overall performance [33]) of the PERSIANN-CDR in detecting extreme precipitation, and does this product reproduce linear trends of extreme precipitation? Particularly, the latter issue has been paid more and more attention in recent years (e.g., [34,35]) because the assessments regarding precipitation trends are the necessary foundation on which to accurately explore precipitation long term changes, especially for the regions with limited and even no observations. Therefore, this study aimed to: (1) comprehensively validate the PERSIANN-CDR performance in detecting different extreme precipitation indices (e.g., the precipitation amount-, duration-, and intensity-based indices) over the HRB, based on four validation metrics (i.e., three continuous validation metrics and one overall performance metric), and (2) detect the PERSIANN-CDR capacity to reproduce linear trends of various extreme precipitation indices. Results of this study will serve as a valuable reference for potential users in the HRB and for the PERSIANN-CDR developers to use to improve the algorithm for obtaining a more accurate extreme precipitation product.

2. Data and Methodology

2.1. Study Region and Data

The HRB is located in eastern China between 30–39°N and 111–123°E (Figure 1). It has a drainage area of approximately 33,000 km^2, covering the northern parts of Jiangsu and Anhui, a small part of Hubei, and most of Shandong and Henan. The HRB has a vast plain, with many lakes and depressions, and is moderately mountainous (elevation generally from 1000 to 2000 m above sea level) near the western boundary, mid-eastern part, and Shandong peninsula. A typical semi-humid monsoon climate prevails in this basin, with regional average annual temperature of 14 °C and precipitation of 806 mm.

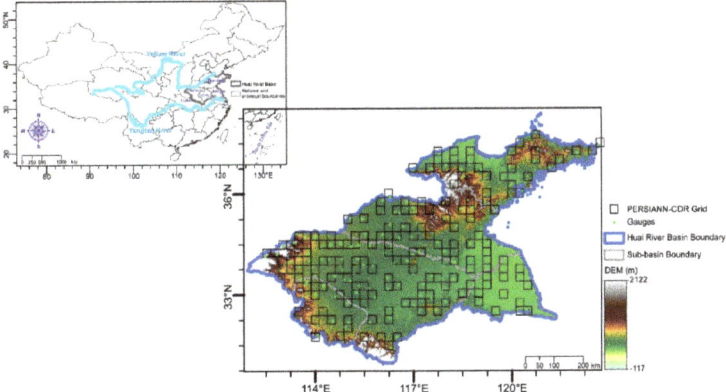

Figure 1. Location of the HRB with gauges and PERSIANN-CDR grids. The digital elevation model (DEM) with a spatial resolution of 90 m is from http://srtm.csi.cgiar.org/ (accessed on 1 January 2021).

The daily PERSIANN-CDR product has near-global (60°S–60°N) coverage with a time span from 1983 to the present and a spatial resolution of 0.25° × 0.25°. It is a new retrospective satellite-based dataset developed by the U.S National Climatic Data Center (NCDC) Climate Data Record program in NOAA [36] and can be downloaded from the U.S NOAA National Centers for Environment Information (NCEI; https://www.ncdc.noaa.

gov/cdr/atmospheric/precipitation-persiann-cdr, accessed on 1 January 2021) and the Centre for Hydrometeorology and Remote Sensing (CHRS) data portal (http://chrsdata.eng.uci.edu, accessed on 1 January 2021). For this evaluation, daily precipitation observed at more than 200 gauges during 1983–2012 were collected from the China Meteorological Administration (CMA). The basic quality issues within the observation precipitation data, e.g., sensors and measurement errors and inherent errors in measurement procedures and methods [37–39], were solved by the CMA. However, it should be noted that data quality issues of missing values and inhomogeneity (e.g., inhomogeneity due to changes in measurement procedures, methods, and locations [36–38]) within observations still remained, and thus we preprocessed the observation data following the procedures below. Firstly, we determined days with missing values for each year and each site. Sites with data available for more than 330 days per year were retained, and missing values of these sites were filled with data from nearby sites by bilinear regression. Subsequently, time series homogeneity was examined with the Pettitt test [40], and the sites with time series not passing the significance test ($p < 0.05$) were removed. Finally, 182 sites remained (Figure 1). To match the PERSIANN-CDR data, we followed Katiraie-Boroujerdy et al. [41] and gridded the sites into grids with a resolution $0.25° \times 0.25°$ (Figure 1). The final observational value for a certain grid was calculated by averaging daily records of the gauge(s) within this grid. Here, the study period is 1983–2012, considering the data availability of both the PERSIANN-CDR and observations.

2.2. Methodology

2.2.1. Extreme Precipitation Index

Due to a lack of a unified definition of extreme event indicators in different regions, further research of global extreme weather and climate events has been hindered to some extent. For addressing this issue, the World Meteorological Organization (WMO) and the World Climate Research Program (WCRP) jointly established the Expert Team on Climate Change Detection and Indices (ETCCDI) in the early 21st century and defined a series of climate indices to study extreme climate change globally and regionally. Since then, the ETCCDI extreme climate indices have been extensively used across the globe [41–46]. In this study, we selected 12 indices to comprehensively evaluate the performance of the PERSIANN-CDR across the HRB. Considering characteristics of extreme precipitation, we categorized the 12 indices into four classes (Table 1), i.e., (1) precipitation amount-based indices, (2) precipitation duration-based indices, (3) precipitation frequency-based indices, and (4) precipitation intensity-based indices.

2.2.2. Validation Metrics

To quantitatively evaluate the performance of PERSIANN-CDR data, we selected a relatively new, widely-used validation metric, the Kling–Gupta Efficiency (KGE; [33]), which can be used to measure overall performance. The equations can be expressed as

$$KGE = 1 - \sqrt{(R-1)^2 + (\beta-1)^2 + (\gamma-1)^2}, \tag{1}$$

$$R = \frac{\sum_{i=1}^{N}(S_i - \mu_s)(O_i - \mu_o)}{\sqrt{\sum_{i=1}^{N}(S_i - \mu_s)^2}\sqrt{\sum_{i=1}^{N}(O_i - \mu_o)^2}}, \tag{2}$$

$$\beta = \frac{\mu_s}{\mu_o}, \tag{3}$$

$$\gamma = \frac{\sigma_s/\mu_s}{\sigma_o/\mu_o}, \tag{4}$$

where S_i is the PERSIANN-CDR precipitation value of the *i*th data pair, and O_i is the observational value. μ_s and μ_o (σ_s and σ_o) are means (standard deviations) of PERSIANN-CDR and observational precipitation, respectively. KGE ranges between—∞ and 1, of which

1 implies a perfect overall performance. R is the correlation coefficient. β measures the average tendency of PERSIANN-CDR precipitation to be larger (i.e., $\beta > 1$) or smaller (i.e., $\beta < 1$) than the observation, with an optimal value of 1. Regarding γ, its optimal value of 1 represents that the PERSIANN-CDR can perfectly reproduce the observational precipitation variability, while values below and above 1, respectively, indicate the underestimated and overestimated variability. After calculating these metrics at each grid with the above equations, their spatial maps were drawn using the ArcGIS 10.2 software package for conveniently comparing the PERSIANN-CDR performance at space.

Table 1. Definitions of the selected 12 extreme precipitation indices.

Class	Name	Definition	Unit
Precipitation amount-based indices	PRCPTOT	Total precipitation on days with precipitation ≥ 1 mm	mm
	R85p	Total precipitation due to events exceeding the 85th percentile of the study period	mm
	R95p	Total precipitation due to events exceeding the 95th percentile of the study period	mm
	R99p	Total precipitation due to events exceeding the 99th percentile of the study period	mm
Precipitation duration-based indices	CDD	Consecutive dry days. Maximum number of consecutive dry days (i.e., when precipitation < 1 mm)	days
	CWD	Consecutive wet days. Maximum number of consecutive wet days (i.e., when precipitation ≥ 1 mm)	days
Precipitation frequency-based indices	R10mm	Number of days with precipitation ≥ 10 mm	days
	R20mm	Number of days with precipitation ≥ 20 mm	days
	Rnnmm	Number of days with precipitation $\geq$ nn mm (nn = 40 mm here)	days
Precipitation intensity-based indices	Rx1day	Maximum 1-day precipitation total	mm/day
	Rx5day	Maximum 5-day precipitation total	mm/(5 days)
	SDII	Simple daily intensity index Total precipitation divided by the number of wet days (i.e., average precipitation of the days with precipitation ≥ 1 mm)	mm/day

3. Results

3.1. Evaluation of Precipitation Amount-Based Indices

Multi-year annual PRCPTOT, R85p, R95p, and R99p from observational precipitation were generally characterized by a decrease from southeastern to northwestern, with the HRB means of 812.60 mm, 441.15 mm, 233.68 mm, and 76.64 mm, respectively (Figure 2a1–a4). Overall, the PERSIANN-CDR could capture a similar spatial distribution for each amount-based index, with spatial Rs of 0.94 for PRCPTOT, 0.92 for R85p, 0.89 for R95p, and 0.81 for R99p (Figure 2b1–b4). Despite that, evident differences in the climatological values of these amount-based indices existed between observation and PERSIANN-CDR (Figure 2b1–b4); the HRB $\beta > 1.0$ indicated that the PERSIANN-CDR overestimated the climatological values of the amount-based indices. Meanwhile, the spatial variabilities of the climatological values were all overestimated, with HRB γ values of 1.40, 1.32, 1.45, and 1.56 for PRCPTOT, R85p, R95p and R99p, respectively. To have an integrative consideration of β, γ, and R, the PERSIANN-CDR showed high (i.e., $KGEs \geq 0.38$) performance overall in spatially representing the climatological value of each amount-based index, especially for R95p with a KGE of 0.58.

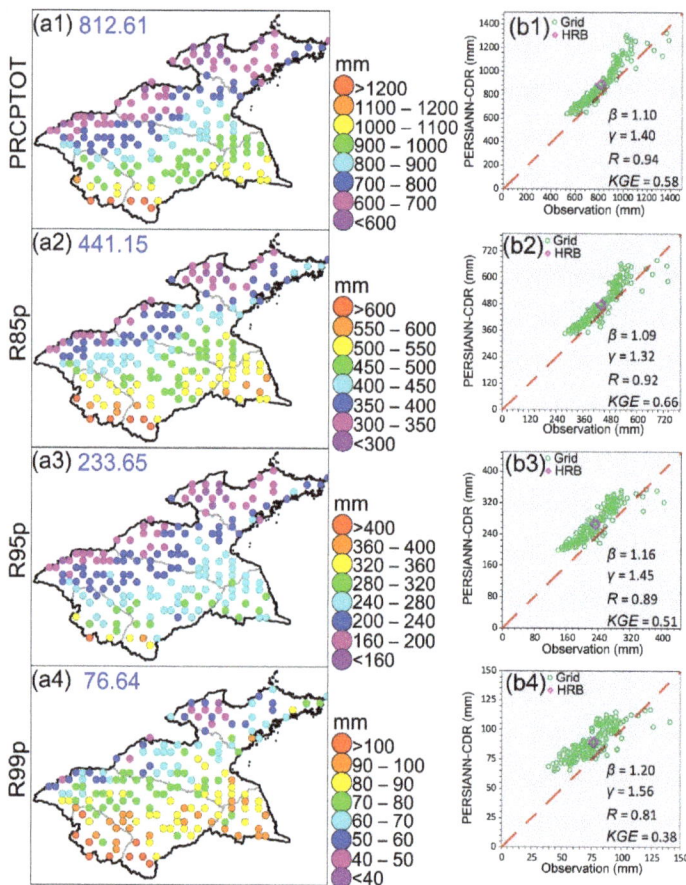

Figure 2. Spatial patterns of multi-year annual means of observational amount-based indices and the scatterplots between observation and PERSIANN-CDR. **a1–a4** (**b1–b4**) are for PRCPTOT, R85p, R95p, and R99p, respectively. In **a1–a4**, the blue numbers represent the HRB mean for a given extreme precipitation index. The red dashed line within **b1–b4** is the 1:1 line.

Generally, each of the four amount-based indices was differently overestimated at > 90% of grids by the PERSIANN-CDR (Figure 3a1–a4). Larger overestimations (i.e., β > 1.4) for PRCPTOT and R85p were mainly located in the southern part (Figure 3a1,a2), while for R95p and R99p larger overestimations were widely distributed across the HRB except for small part of southern HRB (Figure 3a3,a4). For PRCPTOT (R85p), γ values were between 1.0 and 1.2 at >90% of grids, corresponding to overestimated temporal variability; generally, in southwestern and easternmost HRB, temporal variability at <10% of grids was underestimated (Figure 3b1,b2). Regarding R95p and R99p (Figure 3b3,b4), overestimated temporal variabilities existed at >90% of grids, of which >30% of grids had γ > 1.2, mainly in the central-northern part for R95p and in the northern and southeastern parts for R99p. Checking temporal Rs for PRCPTOT at all the grids (Figure 3c1), the values were all >0.50, with 81% of grids showing Rs > 0.70 widely distributed across the HRB. As for R85p (Figure 3c2), most (>85%) grids showed temporal Rs > 0.50, especially for western, southeastern, and northeastern HRB, with temporal Rs > 0.70, while it was noted that there were still some grids with Rs < 0.40 sporadically in the central-northern part. Seen in Figure 3c3, 50% of grids showed Rs > 0.50 for R95p, accompanied by <10% of grids with Rs > 0.70 in southwestern HRB; of the remaining grids, their corresponding Rs < 0.2 indicated that the PERSIANN-CDR had much limited ability in reproducing temporal fluctuations of R95p. Figure 3c4 illustrates that the PERSIANN-CDR could capture temporal fluctuations of R99p at only 15% of grids, mainly in western HRB; moreover, negative Rs in northeastern HRB suggested that the product had no capacity in reproducing temporal fluctuations of R99p. At > 90% of grids, $KGEs$ for both PCPTOT and R85p were >0.20, especially in central-northern HRB, with $KGEs$ > 0.40 indicating better overall performance (Figure 3d1,d2). For R95p (Figure 3d3), there existed 66% of grids with $KGEs$ > 0.2, particularly those in the southern part with $KGEs$ > 0.40, whereas in the northern part around 30% of grids with $KGEs$ < 0.20 showed limited overall performance for representing R95p. Except for the 16% of grids in the southwestern part with $KGEs$ between 0.2 and 0.4, the PERSIANN-CDR lacked the ability to represent R99p over the remaining grids (Figure 3d4).

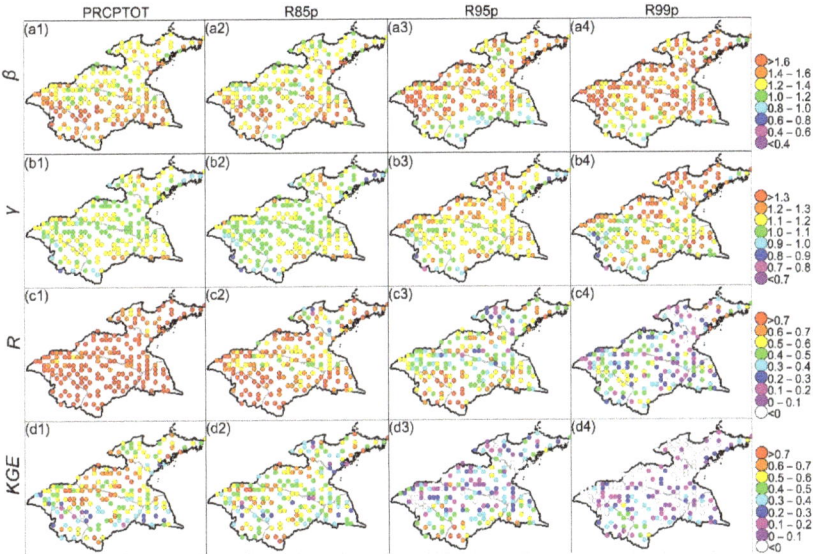

Figure 3. Spatial patterns of different validation metrics for the amount-based indices.

At space, the observational PRCPTOT, R85p, R95p, and R99p trends had a similar distribution, i.e., decreased over western and southeastern parts, but increased in other regions, with the HRB trends of 4.17 mm/yr, 3.68mm/yr, 3.13 mm/yr, and 1.69mm/yr, respectively (Figure 4a1–a4). Moreover, the percentage of the grids with positive trends for each index was always $\geq$ 56%. As shown in Figure 4b1–b4, each of the PERSIANN-CDR amount-based indices corresponded to underestimated trends at most (>50%) grids; for the regional mean, the HRB β < 0.5 suggested that the PERSIANN-CDR seriously underestimated the trends of these amount-based indices, especially for the PRCPTOT with opposite changes (i.e., $\beta = -0.18$) between the observation and the PERSIANN-CDR. Except for PRCPTOT, the spatial variabilities of R85p, R95p, and R99p trends were overestimated with the HRB γ values >1.00. The PERSIANN-CDR showed a moderate performance (spatial R = 0.39) in producing spatial patterns of PRCPTOT trends, but much limited capacity (spatial Rs < 0.20) existed for the other three indices. Based on KGE, it is evident that the PERSIANN-CDR had no ability (i.e., $KGEs$ < 0) to present the trends of these amount-based indices.

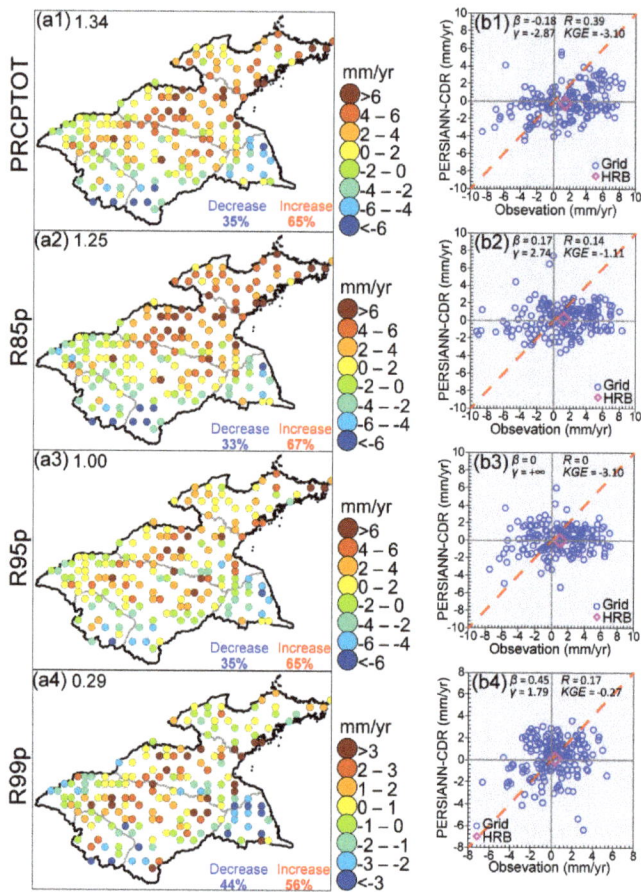

Figure 4. Spatial patterns of the temporal trends of the observational amount-based indices (**a1–a4**), and the scatterplots between observation and PERSIANN-CDR. **a1–a4** (**b1–b4**) are for PRCPTOT, R85p, R95p, and R99p, respectively. In **a1–a4**, the black numbers represent the HRB trends of the observational amount-based indices, while the blue (red) numbers indicate grid percentages with increasing (decreasing) trend across the HRB. The red dashed line in **b1–b4** is the 1:1 line.

3.2. Evaluation of Precipitation Duration-Based Indices

For the HRB, the observational multi-year annual means of CDD and CWD were 45.22 days and 5.08 days, respectively (Figure 5a1,a2), corresponding to spatial distributions of a decrease from northwest to southeast and an increase from northwest to southeast. Based on spatial Rs of 0.86 for CDD and 0.71 for CWD (Figure 5b1,b2), the PERSIANN-CDR better detected spatial distributions of climatological characteristics of these two duration-based indices. It is evident that for the HRB, the PERSIANN-CDR seriously underestimated and overestimated magnitudes of climatological CDD and CWD values, respectively, with β values of 0.68 and 1.95 (Figure 5b1,b2). For spatial variability, larger overestimation existed for CDD with the HRB γ of 1.44, while CWD corresponded to a slight underestimation ($\gamma = 0.98$). In terms of KGE, this PERSIANN-CDR had no ability to represent CWD, while better overall performance ($KGE = 0.44$) existed for CDD (Figure 5b1,b2).

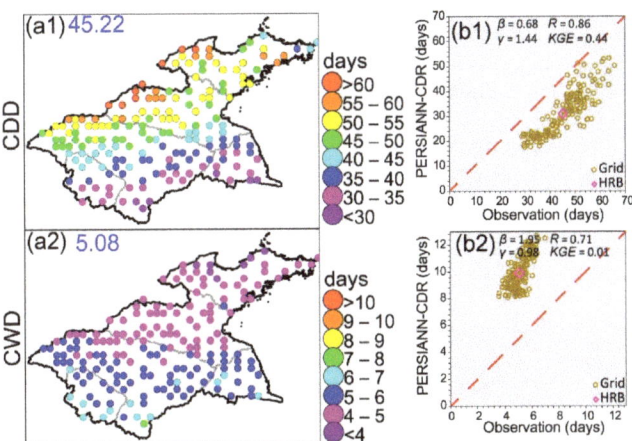

Figure 5. Spatial patterns of multi-year annual means of observational duration-based indices and the scatterplots between observation and PERSIANN-CDR. **a1,a2** (**b1,b2**) are for CDD and CWD, respectively. In **a1,a2**, the blue numbers represent the HRB mean for a given extreme precipitation index. The red dashed line in **b1,b2** is the 1:1 line.

At all the grids, CDD were underestimated ($\beta < 1.00$), followed by > 95% of grids with larger underestimations ($\beta < 0.80$) (Figure 6a1). Conversely, the PERSIANN-CDR much overestimated CWD ($\beta > 1.40$) across the HRB (Figure 6a2). Based on γ, temporal variabilities of CDD were underestimated at > 90% of grids (Figure 6b1), and larger underestimations ($\gamma < 0.9$) mainly appeared in western HRB, followed by some grids with overestimations ($\gamma > 1.00$), mainly in some parts of eastern HRB. For CWD (Figure 6b2), overestimations (underestimations) of temporal variabilities occurred at 25% (75%) of grids but were characterized by sporadic distribution across the study region. Regarding CDD (Figure 6c1), the PERSIANN-CDR had strong ability ($R > 0.50$) to represent temporal fluctuations at 40% of grids in northwestern HRB, but moderate or limited ability at other grids. Except for only 5% of grids with a certain capacity, the PERSIANN-CDR had limited or no capacity ($R < 0.20$) in reproducing temporal fluctuations of CWD across the HRB (Figure 6c2). Seen in Figure 6d1, the PERSIANN-CDR had the ability ($KGE > 0.30$) to represent CDD at >60% of grids in northern HRB, followed by no ability, mainly in southern HRB. Smaller (near to 0) and negative KGEs at all the grids suggested the PERSIANN-CDR had no ability in capturing CWD across the HRB (Figure 6d2).

In view of observations, the two duration-based indices for the HRB differently increased, with a rate of 0.24 days/yr for CDD and 0.02 days/yr for CWD (Figure 7a1,a2). Spatially, the positive trends of the observational CDD occurred at 84% of grids, followed by decreasing trends at 16% of grids in central-northern and southwestern parts (Figure 7a1). There existed >30% of grids with decreased CWD, generally in western HRB, while increased CWD was widely distributed across eastern HRB, with a grid percentage around 70% (Figure 7a2). For the HRB, the CDD trends were overestimated by the PERSIANN-CDR, with β of 1.20 (Figure 7b1), while the product seriously underestimated ($\beta = 0.12$) the CWD trends (Figure 7b2). In terms of γ, the PERSIANN-CDR overestimated spatial variabilities of both CDD and CWD trends, especially for CWD, with a serious overestimation ($\gamma = 10.58$) (Figure 7b1,b2). Overall, there was no ability ($R < 0.10$) for the PERSIANN-CDR to produce spatial patterns of the trends of the duration-based indices, accompanied by no KGE-based ability (KGE near to 0 and even < 0) (Figure 7b1,b2).

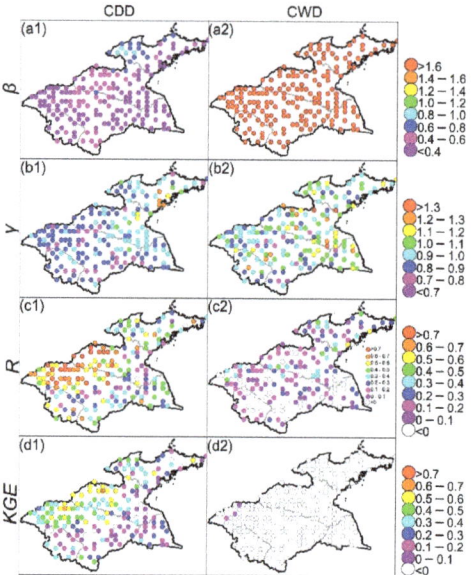

Figure 6. Spatial patterns of different validation metrics for the duration-based indices.

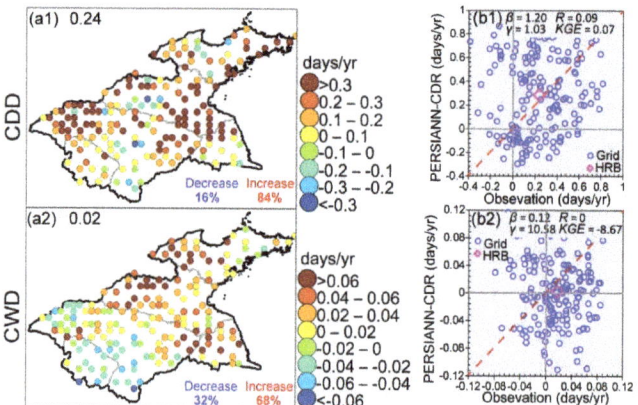

Figure 7. Spatial patterns of the temporal trends of the observational duration-based indices and the scatterplots between observation and PERSIANN-CDR. **a1,a2** (**b1,b2**) are for CDD and CWD, respectively. In **a1,a2**, the black numbers represent the HRB trends of the observational duration-based indices, while the blue (red) numbers indicate grid percentages with increasing (decreasing) trend across the HRB. The red dashed line in **b1,b2** is the 1:1 line.

3.3. Evaluation of Precipitation Frequency-Based Indices

Multi-year annual R10mm, R20mm, and Rnnmm from observational precipitation were characterized by a decrease from northwest to southeast, with the HRB means of 22.93 days, 11.73 days, and 8.86 days, respectively (Figure 8a1–a3). Overall, the PERSIANN-CDR could better capture spatial distributions of climatological R10mm, R20mm, and Rnnmm, with spatial Rs of 0.96, 0.91, and 0.90, respectively (Figure 8b1–b3). For the HRB, magnitudes and spatial variabilities for climatological value of each frequency index were differently underestimated and overestimated by the PERSIANN-CDR, respectively (Figure 8b1–b3). Specifically, the PERSIANN-CDR showed the largest Rnnmm underestimation in magnitude (spatial variability) with the HRB β (γ) of 0.68 (1.30) (Figure 8b1–b3). Based on *KGE*, this product had better overall performance (i.e., *KGE* > 0.55) in representing the three frequency-based indices, particularly for R10mm and R20mm, with *KGEs* > 0.60 (Figure 8b1–b3).

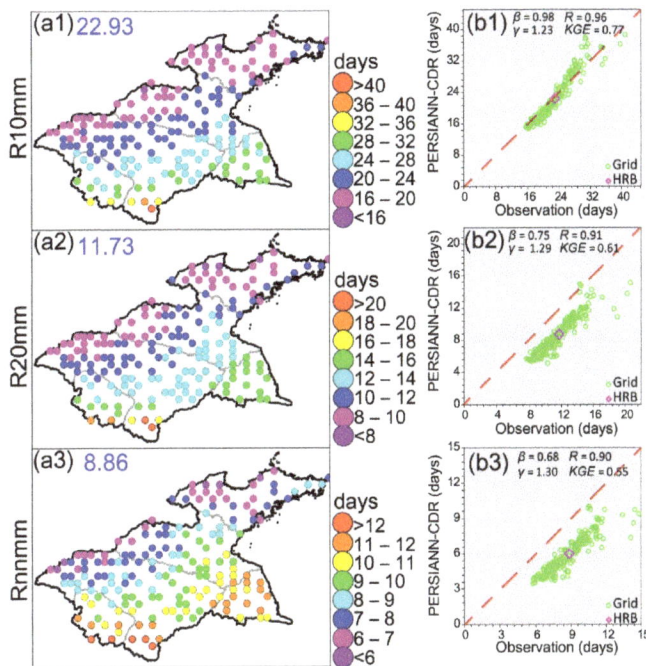

Figure 8. Spatial patterns of multi-year annual means of observational frequency-based indices and the scatterplots between observation and PERSIANN-CDR. **a1–a3** (**b1–b3**) are for R10mm, R20mm, and Rnnmm, respectively. In **a1–a3**, the blue numbers represent the HRB mean for a given extreme precipitation index. The red dashed line in **b1–b3** is the 1:1 line.

Seen in Figure 9a1, except for 4% of grids in the southern part with smaller overestimations (β between 1.00 and 1.10), the PERSIANN-CDR differently underestimated R10mm at the remaining grids. Regarding R20mm and Rnnmm, the underestimations (β < 1.00) occurred at an overwhelming majority (>98%) of grids, of which > 80% of grids corresponded to larger underestimations (β < 0.60) (Figure 9a2,a3). Based on γ, temporal variabilities of the three frequency-based indices were differently underestimated at >75% of grids (Figure 9b1–b3); larger underestimations (γ < 0.8) for R20mm in northern HRB and for Rnnmm in northern and central-southern parts (Figure 9b2,b3). Moreover, there were some grids with overestimated temporal variabilities (γ > 1.00) of the frequency-based indices, e.g., R10mm and R20mm at >15% of grids, generally in southern HRB (Figure 9b1,b2). It is evident that the PERSIANN-CDR had strong ability (R > 0.50) to represent temporal fluctuations of R10mm at 89% of grids, which were widely distributed across the HRB; for R20mm and Rnnmm, there existed >50% of grids with R > 0.50, mainly in southern HRB (Figure 9c1–c3). Obviously, the PERSIANN-CDR exhibited a better overall performance (*KGE* > 0.40) in detecting R10mm at all the grids (Figure 9d1). Except for <15% of grids, generally in northwestern and northeastern HRB with no estimation ability, southern HRB corresponded to *KGEs* > 0.20 for Rn20mm and Rnnmm (Figure 9d2,d3), particularly for most grids of southern HRB, with *KGEs* > 0.5.

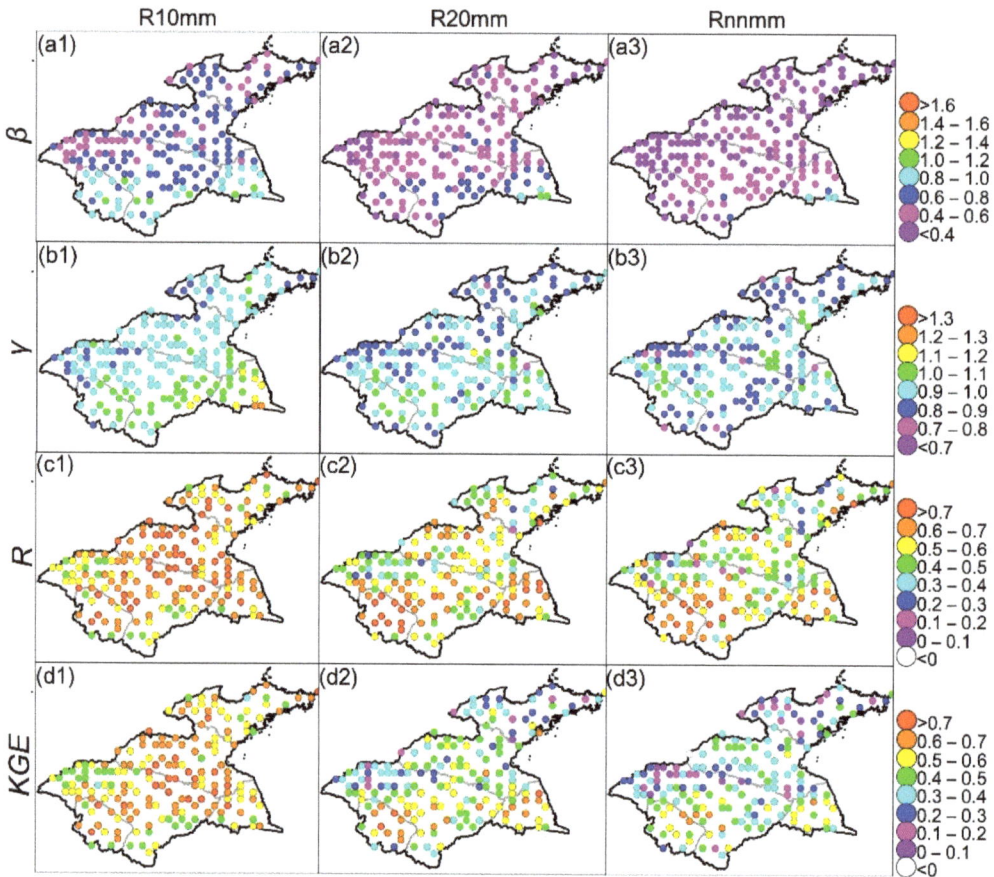

Figure 9. Spatial patterns of different validation metrics for the frequency-based indices.

As shown in Figure 10a1–a3, the HRB R10mm, R20mm, and Rnnmm increased by 0.03 days/yr, 0.03 days/yr, and 0.02 days/yr, respectively. At space, the trends of the observational frequency-based indices generally had a decrease in the western part and an increase in eastern parts; moreover, there were always ≥ 65% of grids with positive trends for the three indices. For the HRB, the PERSIANN-CDR seriously underestimated ($\beta < 0.50$) the trends of all the frequency-based indices, and even for R20mm and Rnnmm, the PERSIANN-CDR showed the opposite trends (Figure 10b1–b3). The metric of γ suggested that this product overestimated spatial variabilities of R10mm (Figure 10b1). There was no ability ($R \leq 0.11$) for the PERSIANN-CDR to produce spatial patterns of R10mm, R20mm, and Rnnmm trends, accompanied by no KGE-based ability (KGE < 0) (Figure 10b1–b3).

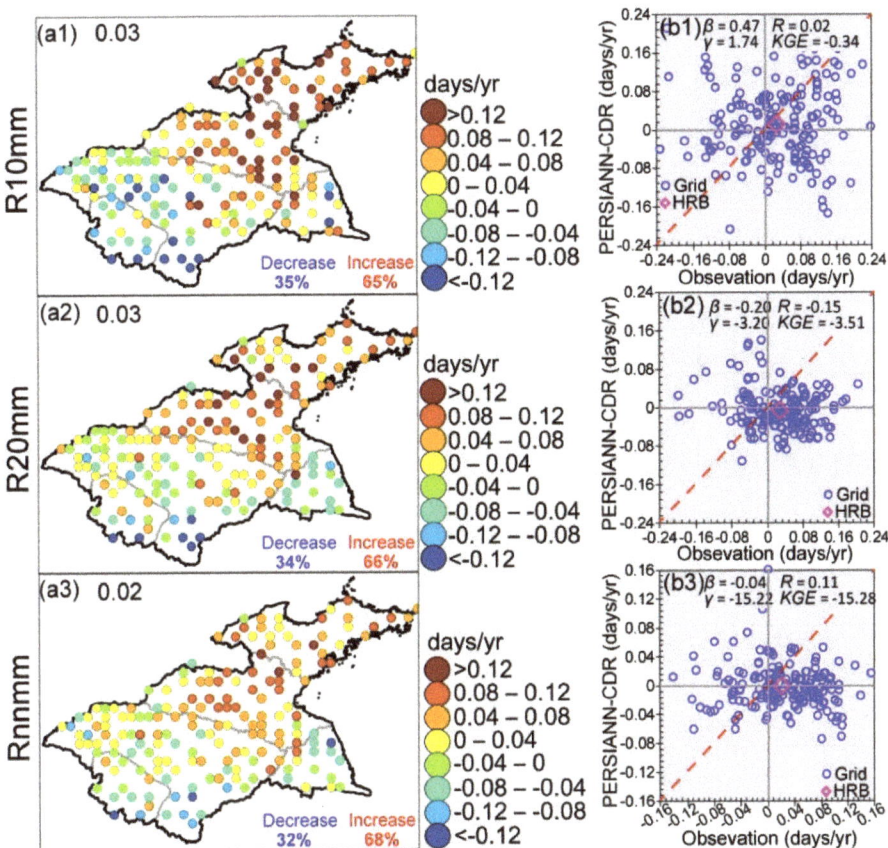

Figure 10. Spatial patterns of the temporal trends of the observational frequency-based indices and the scatterplots between observation and PERSIANN-CDR. **a1–a3** (**b1–b3**) are for R10mm, R20mm, and Rnnmm, respectively. In **a1–a3**, the black numbers represent the HRB trends of the observational frequency-based indices, while the blue (red) numbers indicate grid percentages with increasing (decreasing) trend across the HRB. The red dashed line in **b1–b3** is the 1:1 line.

3.4. Evaluation of Precipitation Intensity-Based Indices

For Rx1day, Rx5day, and SDII, the observational multi-year annual means were 96.05 mm/day, 148.97 mm/(5 days), and 12.99 mm/day for the HRB, respectively, generally characterized by an increase from the northwest to southeast (Figure 11a1–a3). The spatial R_s of 0.25 for Rx1day, 0.38 for Rx5day, and 0.52 for SDII indicated that the PERSIANN-CDR could reproduce spatial patterns of climatological characteristics of the intensity-based indices (Figure 11b1–b3). The HRB β values < 0.80 for the intensity-based indices suggested that the three indices were underestimated by the PERSIANN-CDR, especially for SDII ($\beta = 0.49$), followed by R1xday ($\beta = 0.64$). For Rx1day and Rx5day, the HRB γ values < 1.0 indicated that spatial variabilities of the two PERSIANN-CDR intensity-based indices were smaller than the observations (Figure 11b1,b2), followed by SDII with γ of 1.37. Based on *KGE*, this product had a moderate overall performance (*KGE* > 0.20) in representing the three intensity-based indices.

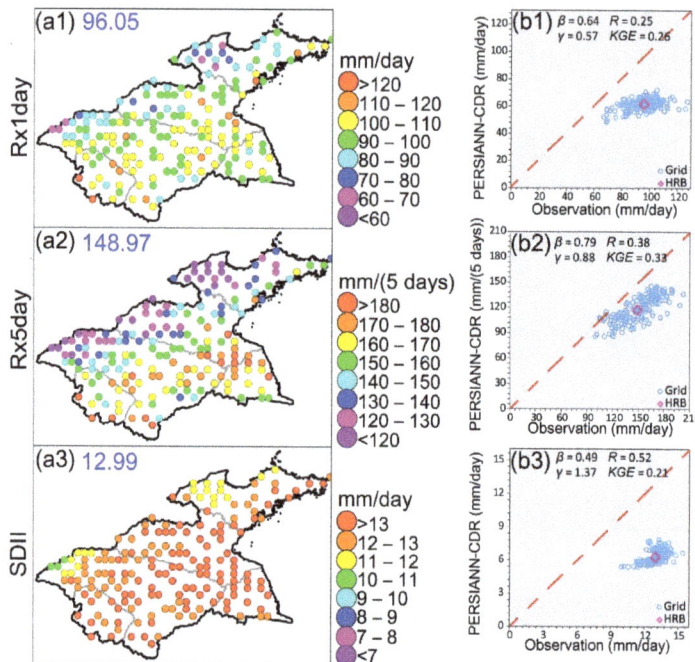

Figure 11. Spatial patterns of multi-year annual means of the observational intensity-based indices and the scatterplots between observation and PERSIANN-CDR. **a1–a3** (**b1–b3**) are for Rx1day, Rx5day, and SDII, respectively. In **a1–a3**, the blue numbers represent the HRB mean for a given extreme precipitation index. The red dashed line in **b1–b3** is the 1:1 line.

In general, the intensity-based indices were underestimated by PERSIANN-CDR except for only 3% of grids with slight overestimations (β between 1.00 and 1.20) for Rx5day in the northeastern part (Figure 12a1–a3). There were more than 80% of grids with overestimated temporal variabilities for the three indices, especially in northwestern and central-eastern HRB, with $\gamma > 1.40$ for Rx1day and Rx5day, and in northwestern and southeastern HRB, with $\gamma > 1.30$ for SDII (Figure 12b1–b3). The PERSIANN-CDR had strong or moderate ability ($R > 0.30$) in detecting temporal fluctuations of R1xday at 39% of grids, but ability was sporadically distributed across the HRB (Figure 12c1). For Rx5day (Figure 12c2), there existed 68% of grids with $R > 0.30$, of which 30% of grids with better R-based performance ($R > 0.50$) were generally in southern HRB; moreover, the PERSIANN-CDR showed limited ($R < 0.30$) or no ability in reproducing temporal variability, particularly in the northern part with $R < 0.20$ and even negative. For SDII (Figure 12c3), >90% of grids with $R > 0.30$ suggested that the PERSIANN-CDR had the ability to reproduce temporal variability across the HRB, especially for western and southeastern parts, with better R-based performance ($R > 0.50$). Spatially, the product had the ability ($KGE > 0.20$) to represent Rx1day at 28% of grids, mainly in middle HRB, but no ability at 72% of grids (Figure 9d1). With exception of 40% of grids having no ability, generally in northern HRB, the PERSIANN-CDR corresponded to a better overall performance for Rx5day across southern HRB, especially in the southeastern part, with $KGE > 0.40$ (Figure 9d2). The PERSIANN-CDR exhibited a certain overall performance ($KGE > 0.20$) in detecting SDII at 59% of grids, followed by 41% of grids with limited and even no ability (Figure 12d3).

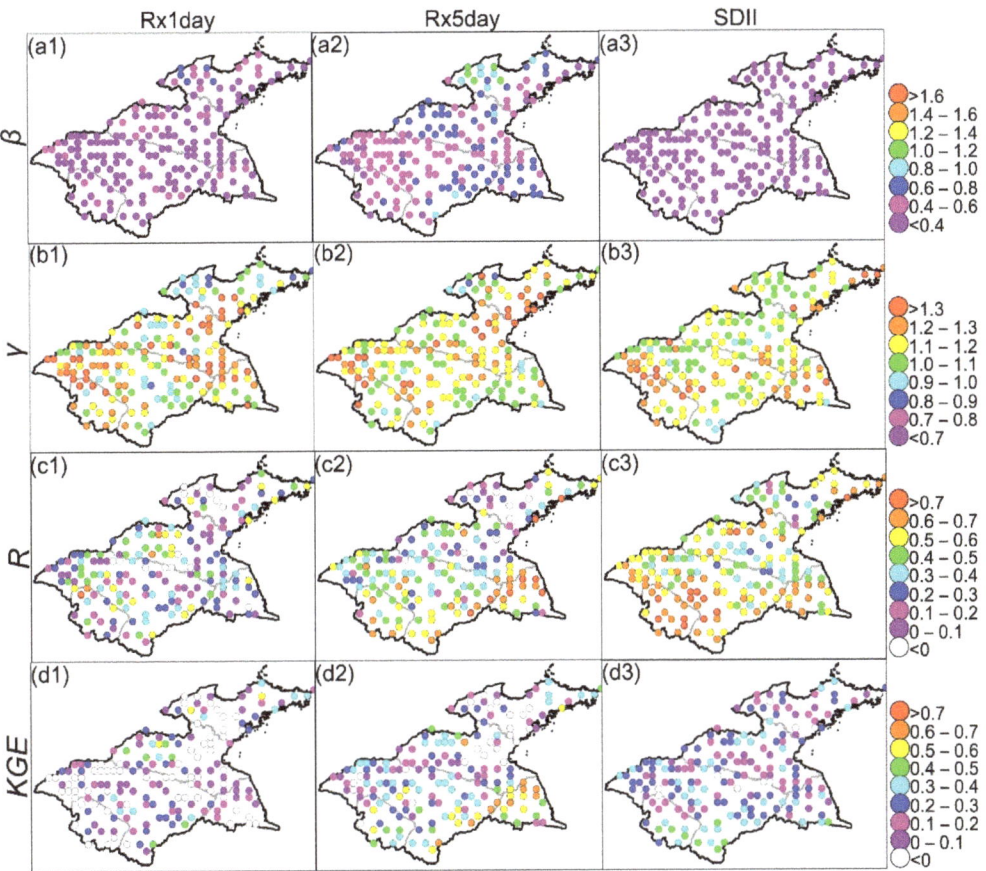

Figure 12. Spatial patterns of different validation metrics for the intensity-based indices.

All the observational precipitation intensity-based indices for the HRB increased but at different rates, i.e., 0.23 mm/(day yr) for Rx1day, 0.76 mm/(5 days yr) for Rx5day, and 0.03 mm/(day yr) for SDII (Figure 13a1–a3). Generally, the measured Rx1day increased at most (59%) grids, followed by 41% of grids with decreased Rx1day, while the spatial distribution was scattered (Figure 13a1). For the observational Rx5day, 74% of grids corresponded to an increase, particularly in western HRB (excluding southwestern part) with a rate 1.00 mm/(5 days yr), while the remaining grids, generally in the southwestern and southeastern parts, showed different reductions (Figure 13a2). There were 34% of grids with decreased SDII, mainly in the southwestern and southeastern parts, followed by increases at the remaining grids (Figure 13a3). Broadly, the HRB β values for the intensity-based indices were all ≤ 0.52, suggesting underestimated trends by the PERSIANN-CDR, especially for Rx5day trends, with many underestimations ($\beta = 0.20$) (Figure 13b1–b3). In terms of the HRB γ, the PERSIANN-CDR underestimated spatial variabilities ($\gamma < 0.90$) for the HRB Rx1day and SDII trends but overestimated ($\gamma < 1.31$) Rx5day trends (Figure 10b1–b3). The PERSIANN-CDR had a certain R-based performance (spatial R around 0.20 or > 0.30) in producing spatial patterns of these indices' trends (Figure 13b1–b3). There was no ability ($KGE < 0.10$) for the PERSIANN-CDR to represent these trends (Figure 10b1–b3).

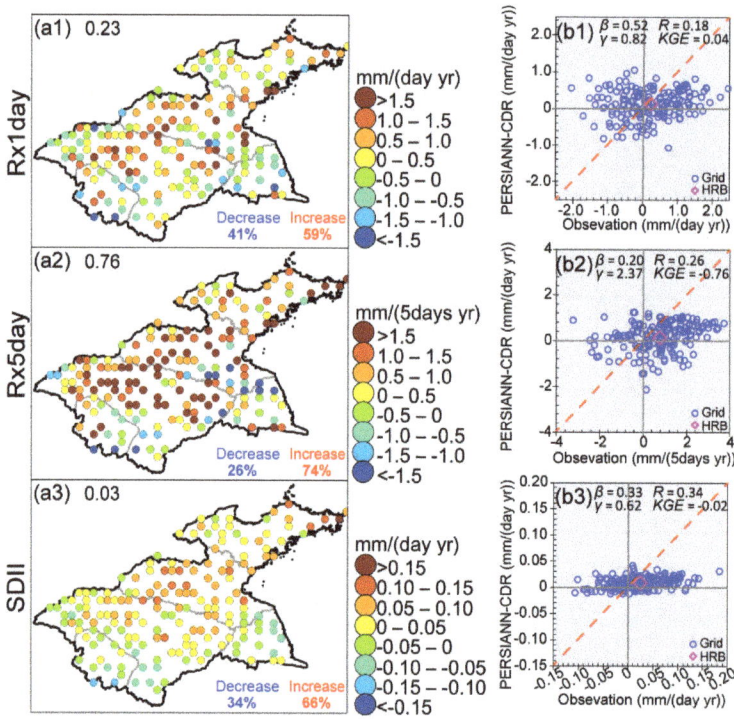

Figure 13. Spatial patterns of the temporal trends of the observational intensity-based indices and the scatterplots between observation and PERSIANN-CDR. **a1–a3** (**b1–b3**) are for Rx1day, Rx5day, and SDII, respectively. In **a1–a3**, the black numbers represent the HRB trends of the observational intensity-based indices, while the blue (red) numbers indicate grid percentages with increasing (decreasing) trend across the HRB. The red dashed line in **b1–b3** is the 1:1 line.

4. Conclusions and Discussion

Attempts to validate various satellite-based precipitation products' capacity in representing precipitation characteristics from different perspectives have been widely conducted all over the world. However, information about their capacity in detecting extreme precipitation and related changes (i.e., linear trends) is scarce. As a result, we collected daily observations from 182 gauges across the HRB during 1983–2012 and examined the PERSIANN-CDR capacity to represent precipitation amount- (PRCPTOT, R85p, R95p, and R99p), duration- (CDD and CWD), frequency- (R10mm, R20mm, and Rnnmm), and intensity-based (Rx1day, R5xday, and SDII) indices and their linear trends. The conclusions can be summarized as follows.

(1) Validation for amount-based indices. Overall, the PERSIANN-CDR could well capture climatological characteristics of the amount-based indices, but with overestimations in magnitudes and spatial variabilities for the HRB. At most grids, both magnitudes and temporal variabilities of each amount-based index were differently overestimated. Generally, the PERSIANN-CDR had better R- and KGE-based performance in producing the amount-based indices (excluding R99p) across the HRB. The linear trend of each amount-based index was underestimated at most grids. Except for PRCPTOT, overestimations (limited capacity) existed for spatial variabilities (spatial patterns) of the other indices' trends. Broadly, the PERSIANN-CDR had no KGE-based ability to present the trends of the four indices.

(2) Validation for duration-based indices. Though the PERSIANN-CDR better detected spatial distributions of climatological characteristics of the duration-based indices,

it underestimated and overestimated climatological values of the HRB CDD and CWD, respectively. For spatial variabilities, overestimations existed for the climatological CDD, but underestimations for the climatological CWD. The PERSIANN-CDR showed no *KGE*-based ability and better overall performance in representing the climatological CWD and CDD, respectively. Over most of the HRB, CDD (CWD) were underestimated (overestimated), with underestimations of temporal variabilities. For most grids, the PERSIANN-CDR had strong and moderate ability to represent temporal fluctuations of CDD, with moderate *KGE*-based performance; however, the opposite results were found for CWD. The HRB CDD and CWD trends were overestimated and underestimated, respectively, followed by overestimated spatial variabilities. Overall, the PERSIANN-CDR had no *R*-based ability in producing spatial patterns of the trends of the duration-based indices, accompanied with no *KGE*-based ability.

(3) Validation for frequency-based indices. The PERSIANN-CDR could better capture spatial distributions of climatological R10mm, R20mm, and Rnnmm, with better *KGE*-based performance. For the HRB, magnitudes and spatial variabilities for the climatological values of each frequency-based index were differently underestimated and overestimated, respectively. Across the HRB, the R10mm underestimations and the R20mm and Rnnmm overestimations were widely distributed. For temporal variabilities, all the frequency-based indices were underestimated at most grids. In general, the PERSIANN-CDR had strong ability to represent temporal fluctuations of the three indices across the HRB. Moreover, there existed *KGE*-based ability for this product to detect these indices, especially for R10mm, with a better overall performance. For the HRB, the PERSIANN-CDR seriously underestimated the trends of the frequency-based indices and overestimated spatial variabilities of R10mm. No *R*-based ability and *KGE*-based ability existed for the PERSIANN-CDR to capture the trends of the frequency-based indices.

(4) Validation for intensity-based indices. The PERSIANN-CDR had ability to reproduce spatial patterns of climatological characteristics of the intensity-based indices, but with underestimated magnitudes. Except for SDII, the other two indices both corresponded to different underestimations in spatial variabilities of climatological values. This product had a moderate *KGE*-based performance in representing climatological values of the intensity-based indices. Across the HRB, the intensity-based indices were generally underestimated, but their temporal variabilities were overestimated. With the exception of R1xday, the PERSIANN-CDR exhibited ability to reproduce temporal variabilities of Rx5day and SDII across most of the HRB. No *KGE*-based ability was detected at most grids for Rx1day, while the PERSIANN-CDR corresponded to a better and a certain *KGE*-based performance for Rx5day and SDII at most grids, respectively. As for the trends, underestimations existed in magnitudes and spatial variabilities for the intensity-based indices (except for Rx5day). The PERSIANN-CDR showed a certain *R*-based ability in reproducing spatial patterns of these indices' trends, but no *KGE*-based abilities existed.

A comprehensive assessment of the PERSIANN-CDR extreme precipitation over the HRB was conducted by comparing here with gauge measurements. However, it should be noted that there existed some issues—e.g., mismatch in spatial scale between point-scale gauge and areal satellite precipitation, inherent uncertainties for gauge observations (including calibration flaws, wind-related undercatch, wetting-evaporation losses, etc.), and inhomogeneity of observations—influencing the confidence level of our findings [39,47–52]. Because of precipitation with large variability at a small spatial extent, a sparse gauge network is difficult to use to fully detect precipitating processes at a given PERSIANN-CDR grid. Therefore, to minimize the related uncertainties into validation results, a sufficient number of gauges should be collected [47]. Commonly, gauges have flaws in calibration, consequently resulting in measured values with uncertainties. For instance, some studies have stated that calibration flaws tended to underestimate gauge observations, particularly for greater rainfall intensities [48]. Under wind-related undercatch effect, the catch efficiency of gauges becomes lower, more or less, mainly due to raindrops missing the funnel or falling at an inclination. As a result, the gauged-recorded precipitation is often smaller than the true values, and underestimations are closely associated with ambient wind speed, raindrop size distribution, and gauge design [49]. Moreover, the gauge values are likely to be underestimated because of evaporation from water adhering to the inside walls of the gauges (i.e., wetting losses) and exposure of the water surface within a gauge to atmosphere (i.e., evaporation losses) [50]. Simply, these influential factors of gauge measurements have an aggregate impact of underestimating gauge precipitation, which then propagate impact into our results [51]. In this study, although gauges with inhomogeneous observations were removed with the Pettitt test (a better method to examine observations' homogeneity when lacking meta-data for gauges; [40]), no guarantee shows that the records at the remaining gauges were all homogenous, potentially weakening the confidence level of this study.

Regardless, our study provides some significant reference data for PERSIANN-CDR developers and potential users in the HRB and other regions. For example, the different capacity of PERSIANN-CDR to detect various extreme precipitation indices suggests that PERSIANN-CDR developers might try to develop specific algorithms and/or correction procedures for increasing a certain validation metric-based performance; for potential users, some PERSIANN-CDR extreme precipitation indices (e.g., CWD, Rx1day, and Rx5day) with poor performance should be excluded from use. The poor performance of PERSIANN-CDR for detecting linear trends of all the selected indices implies that more effort should be devoted by the developers to improving PERSIANN-CDR's abilities; moreover, more attention should be paid by potential users of PERSIANN-CDR when conducting studies of long-term changes in extreme precipitation.

Author Contributions: S.S., J.W., and W.S. conceived and designed this study. S.S., J.W., W.S., and R.C. were the main authors, whose work included data collection and analysis, interpretation of results, and manuscript preparation. G.W. played a supervisory role. S.S., J.W., W.S., and R.C. contributed by processing data and providing rain gauge observations. All authors discussed the results and revised the manuscript. All authors have read and agreed to the published version of the manuscript.

Funding: This work was jointly supported by the National Key Research and Development Program of China (Grant NOs. 2018YFC1507101 and 2017YFA0603701), the National Natural Science Foundation of China (Grant NOs. 42075189, 41605042, and 41875094), and the Qinglan Project of Jiangsu Province of China.

Acknowledgments: PERSIANN-CDR daily precipitation data were downloaded from the Centre for Hydrometeorology and Remote Sensing (CHRS) data portal with a website at http://chrsdata.eng.uci.edu (accessed on 1 January 2021), while DEM data of SRTM3 are available from http://srtm.csi.cgiar.org/index.asp (accessed on 1 January 2021). Notably, daily precipitation observations at more than 200 gauge sites are not available to the public, but they can be obtained and used through cooperation with the CMA. We thank all data developers and their managers and funding agencies, whose work and support were essential for obtaining the datasets, without which the analyses conducted in this study would have been impossible. In addition, source code for conducting this study is available from the authors upon request (sun.s@nuist.edu.cn).

Conflicts of Interest: The authors declare no conflict of interest.

References

1. IPCC. *Summary for policymakers. Climate Change 2014: Impacts, Adaptation, and Vulnerability. Part A: Global and Sectoral Aspects. Contribution of Working Group ii to the Fifth Assessment Report of the Intergovernmental Panel on Climate Change*; Field, C.B., Barros, V.R., Dokken, D.J., Mach, K.J., Mastrandrea, M.D., Bilir, T.E., Chatterjee, M., Ebi, K.L., Estrada, Y.O., Genova, R.C., et al., Eds.; Cambridge University Press: Cambridge, UK, 2014; pp. 1–32.
2. Milly, P.; Wetherald, R.; Dunne, K.A.; Delworth, T.L. Increasing risk of great floods in a changing climate. *Nature* **2002**, *415*, 514–517. [CrossRef]

3. Wang, G.L. Lessons learned from protective measures associated with the 2010 Zhouqu debris flow disaster in China. *Nat. Hazards* **2013**, *69*, 1835–1847. [CrossRef]
4. Witze, A. Why extreme rains are gaining strength as the climate warms. *Nature* **2018**, *563*, 458–460. [CrossRef]
5. Amarnath, G.; Yoshimoto, S.; Goto, O.; Fujihara, M.; Smakhtin, V.; Aggarwal, P.K.; Ravan, S. Global Trends in Water-Related Disasters Using Publicly Available Database for Hazard and Risk Assessment. Available online: https://cgspace.cgiar.org/bitstream/handle/10568/93032/H048407.pdf (accessed on 3 August 2020).
6. Lynch, S.L.; Schumacher, R.S. Ensemble-based analysis of the May 2010 extreme rainfall in Tennessee and Kentucky. *Mon. Weather Rev.* **2014**, *142*, 222–239. [CrossRef]
7. Martius, O.; Sodemann, H.; Joos, H.; Pfahl, S.; Winschall, A.; Croci-Maspoli, M.; Sprenger, M.; Wernli, H.; Sedláček, J.; Schemm, S.; et al. The role of upper-level dynamics and surface processes for the Pakistan flood of July 2010. *Q. J. Roy. Meteorol. Soc.* **2013**, *139*, 1780–1797. [CrossRef]
8. Ávila, Á.; Guerrero, F.C.; Escobar, Y.C.; Justino, F. Recent Precipitation Trends and Floods in the Colombian Andes. *Water* **2019**, *11*, 379. [CrossRef]
9. Duan, W.; He, B.; Nover, D.; Fan, J.; Yang, G.; Chen, W.; Meng, H.; Liu, C. Floods and associated socioeconomic damages in China over the last century. *Nat. Hazards* **2016**, *82*, 401–413. [CrossRef]
10. Tustison, B.; Harris, D.; Foufoula-Georgiou, E. Scale issues in verification of precipitation forecasts. *J. Geophysi. Res. Atmos.* **2001**, *106*, 11775–11784. [CrossRef]
11. Alijanian, M.; Rakhshandehroo, G.R.; Mishra, A.K.; Dehghani, M. Evaluation of satellite rainfall climatology using CMORPH, PERSIANN-CDR, PERSIANN, TRMM, MSWEP over Iran. *Int. J. Climatol.* **2017**, *37*, 4896–4914. [CrossRef]
12. Westrick, K.J.; Mass, C.F.; Colle, B.A. The limitations of the WSR-88D radar network for quantitative precipitation measurement over the Coastal Western United States. *Bull. Am. Meteorol. Soc.* **1999**, *80*, 2289–2298. [CrossRef]
13. Young, C.B.; Nelson, B.R.; Bradley, A.A.; Smith, J.A.; Peters-Lidard, C.D.; Kruger, A.; Baeck, M.L. An evaluation of NEXRAD precipitation estimates in complex terrain. *J. Geophysi. Res. Atmos.* **1999**, *104*, 19691–19703. [CrossRef]
14. AghaKouchak, A.; Behrangi, A.; Sorooshian, S.; Hsu, K.; Amitai, E. Evaluation of satellite-retrieved extreme precipitation rates across the central United States. *J. Geophysi. Res. Atmos.* **2011**, *116*. [CrossRef]
15. Huffman, G.J. README for Accessing Experimental Realtime TRMM Multi-Satellite Precipitation Analysis (Tmpart) Data Sets. NASA Tech. Doc. 2015; p. 12. Available online: Ftp://mesoa.gsfc.nasa.gov/pub/trmmdocs/rt/3B4XRT_README.pdf (accessed on 1 January 2020).
16. Joyce, R.J.; Janowiak, J.E.; Arkin, P.A.; Xie, P. CMORPH: A method that produces global precipitation estimates from passive microwave and infrared data at high spatial and temporal resolution. *J. Hydrometeorol.* **2014**, *5*, 487–503. [CrossRef]
17. Maggioni, V.; Sapiano, M.R.P.; Adler, R.F. Estimating uncertainties in high-resolution satellite precipitation products: Systematic or random error? *J. Hydrometeorol.* **2016**, *17*, 1119–1129. [CrossRef]
18. Sorooshian, S.; AghaKouchak, A.; Arkin, P.; Eylander, J.; Foufoula-Georgiou, E.; Harmon, R.; Hendrickx, J.M.H.; Imam, B.; Kuligowski, R.; Skahill, B.; et al. Advanced concepts on remote sensing of precipitation at multiple scales. *Bull. Am. Meteorol. Soc.* **2011**, *92*, 1353–1357. [CrossRef]
19. Huffman, G.J.; Adler, R.F.; Bolvin, D.T.; Gu, G.; Nelkin, E.J.; Bowman, K.P.; Yong, Y.; Stocker, E.F.; Wolff, D.B. The TRMM Multi-satellite Precipitation Analysis (TMPA): Quasi-global, multi-year, combined-sensor precipitation at fine scales. *J. Hydrometeorol.* **2007**, *8*, 38–55. [CrossRef]
20. Funk, C.C.; Peterson, P.J.; Landsfeld, M.F.; Pedreros, D.H.; Verdin, J.P.; Sukla, S.; Husak, G.J.; Rowland, J.D.; Harrison, L.; Hoell, A.; et al. The climate hazards infrared precipitation with stations—A new environmental record for monitoring extremes. *Sci. Data* **2015**, *2*, 150066. [CrossRef]
21. Shen, Y.; Xiong, A.; Wang, Y.; Xie, P. Performance of high-resolution satellite precipitation products over China. *J. Geophysi. Res. Atmos.* **2010**, *115*. [CrossRef]
22. Tan, M.; Ibrahim, A.; Duan, Z.; Cracknell, A.; Chaplot, V. Evaluation of six high-resolution satellite and ground-based precipitation products over Malaysia. *Remote Sens.* **2015**, *7*, 1504–1528. [CrossRef]
23. Liu, X.; Yang, T.; Hsu, K.; Liu, C.; Sorooshian, S. Evaluating the streamflow simulation capability of PERSIANN-CDR daily rainfall products in two river basins on the Tibetan Plateau. *Hydrol. Earth Syst. Sci.* **2017**, *21*, 169–181. [CrossRef]
24. She, D.; Xia, J.; Zhang, Y.; Du, H. The trend analysis and statistical distribution of extreme rainfall events in the Huaihe River Basin in the past 50 years. *Acta Geo. Sin.* **2011**, *66*, 1200–1210. (In Chinese with English Abstract)
25. Bi, B.; Jiao, M.; Li, Z. Contrast analysis of meteorological and hydrological features of extremely heavy rainfall causing severe floods in Huai River Valley. *J. Nanjing Inst. Meteorol.* **2004**, *27*, 577–586. (In Chinese with English Abstract)
26. Wei, F.; Zhang, T. Oscillation characteristics of summer precipitation in the Huaihe River valley and relevant climate background. *Sci. China Earth Sci.* **2009**, *39*, 1360–1374. (In Chinese) [CrossRef]
27. Zhang, J. The 2003 floods in Huai River Basin. *Meteorol. Knowl.* **2003**, *5*, 2–4. (In Chinese)
28. Yin, J.; Yan, D.; Yang, Z.; Yuan, Z.; Yuan, Y.; Zhang, C. Projection of extreme precipitation in the context of climate change in Huang-Huai-Hai region, China. *J. Earth Syst. Sci.* **2016**, *125*, 417–429. [CrossRef]
29. Miao, C.; Ashouri, H.; Hsu, K.-L.; Sorooshian, S.; Duan, Q. Evaluation of the PERSIANN-CDR daily rainfall estimates in capturing the behavior of extreme precipitation events over China. *J. Hydrometeorol.* **2015**, *16*, 1387–1396. [CrossRef]

30. Liu, J.; Xu, Z.; Bai, J.; Peng, D.; Ren, M. Assessment and correction of the PERSIANN-CDR product in the Yarlung Zangbo River Basin, China. *Remote Sens.* **2018**, *10*, 2031. [CrossRef]
31. Liu, J.; Xia, J.; She, D.; Li, L.; Wang, Q.; Zou, L. Evaluation of six satellite-based precipitation products and their ability for capturing characteristics of extreme precipitation events over a climate transition area in China. *Remote Sens.* **2019**, *11*, 1477. [CrossRef]
32. Gao, F.; Zhang, Y.; Chen, Q.; Wang, P.; Yang, P.; Yao, Y.; Cai, W. Comparison of two long-term and high-resolution satellite precipitation datasets in Xinjiang, China. *Atmos. Res.* **2018**, *212*, 150–157. [CrossRef]
33. Gupta, H.V.; Kling, H.; Yilmaz, K.K.; Martinez, G.F. Decomposition of the mean squared error and NSE performance criteria: Implications for improving hydrological modelling. *J. Hydrol.* **2012**, *377*, 80–91. [CrossRef]
34. Sun, S.; Shi, W.; Zhou, S.; Chai, R.; Chen, H.; Wang, G.; Zhou, Y.; Shen, H. Capacity of satellite-based and reanalysis precipitation products in detecting long-term trends across Mainland China. *Remote Sens.* **2020**, *12*, 2902. [CrossRef]
35. Chen, F.; Gao, Y. Evaluation of precipitation trends from high-resolution satellite precipitation products over Mainland China. *Clim. Dyn.* **2018**, *51*, 3311–3331. [CrossRef]
36. Ashouri, H.; Hsu, K.; Sorooshian, S.; Braithwaite, D.K.; Knapp, K.R.; Cecil, L.D.; Nelson, B.R.; Prat, O.P. PERSIANNCDR: Daily precipitation climate data record from multi-satellite observations for hydrological and climate studies. *Bull. Am. Meteorol. Soc.* **2015**, *96*, 69–83. [CrossRef]
37. Estévez, J.; Gavilán, P.; Giráldez, J.V. Guidelines on validation procedures for meteorological data from automatic weather stations. *J. Hydrol.* **2011**, *402*, 144–154. [CrossRef]
38. Gentilucci, M.; Barbieri, M.; Burt, P.; D'Aprile, F. Preliminary data validation and reconstruction of temperature and precipitation in Central Italy. *Geosciences* **2018**, *8*, 202. [CrossRef]
39. Zahumenský, I. *Guidelines on Quality Control Procedures for Data from Automatic Weather Stations*; World Meteorological Organization: Geneva, Switzerland, 2004.
40. Wijngaard, J.B.; Tank, A.M.G.K.; Konnen, G.P. Homogeneity of 20th century European daily temperature and precipitation series. *Int. J. Climatol.* **2003**, *23*, 679–692. [CrossRef]
41. Katiraie-Boroujerdy, P.S.; Nasrollahi, N.; Hsu, K.-L.; Sorooshian, S. Evaluation of satellite-based precipitation estimation over Iran. *J. Arid. Environ.* **2013**, *97*, 205–219. [CrossRef]
42. Alexander, L.V.; Zhang, X.; Peterson, T.C.; Caesar, J.; Klein, T.A.M.G.; Haylock, M.; Collins, D.; Trewin, B.; Rahimzadeh, F.; Tgipour, A.; et al. Global observed changes in daily climate extremes of temperature and precipitation. *J. Geophys. Res. Atmos.* **2006**, *111*. [CrossRef]
43. Kim, Y.H.; Min, S.K.; Zhang, X.B.; Zwiers, F.; Alexander, L.V.; Donat, M.G.; Tung, Y.-S. Attribution of extreme temperature changes during 1951–2010. *Clim. Dyn.* **2016**, *46*, 1769–1782. [CrossRef]
44. Donat, M.G.; Alexander, L.V.; Yang, H.; Durre, I.; Vose, R.; Caesar, J. Global land-based datasets for monitoring climatic extremes. *Bull. Am. Meteorol. Soc.* **2013**, *94*, 997–1006. [CrossRef]
45. Yin, H.; Sun, Y. Detection of anthropogenic influence on fixed threshold indices of extreme temperature. *J. Clim.* **2018**, *31*, 6341–6352. [CrossRef]
46. Cerón, W.L.; Kayano, M.T.; Andreoli, R.V.; Avila-Diaz, R.; Ayes, I.; Freitas, E.D.; Martins, J.A.; Souza, R.A.F. Recent intensification of extreme precipitation events in the La Plata Basin in Southern South America (1981–2018). *Atmos. Res.* **2020**, *249*, 105299. [CrossRef]
47. Villarini, G.; Mandapaka, P.V.; Krajewski, W.F.; Moore, R.J. Rainfall and sampling uncertainties: A rain gauge perspective. *J. Geophysi. Res. Atmos.* **2008**, *113*. [CrossRef]
48. Shedekar, V.S.; King, K.W.; Fausey, N.R.; Soboyejo, A.B.O.; Harmel, R.D.; Brown, L.C. Assessment of measurement errors and dynamic calibration methods for three different tipping bucket rain gauges. *Atmos. Res.* **2016**, *178*, 445–458. [CrossRef]
49. Pollock, M.D.; O'Donnell, G.; Quinn, P.; Dutton, M.; Black, A.; Wilkinson, M.E.; Colli, M.; Stagnaro, M.; Lanza, L.G.; Lewis, E.; et al. Quantifying and mitigating wind-Induced undercatch in rainfall measurements. *Water Resour. Res.* **2018**, *54*, 3863–3875. [CrossRef]
50. Adam, J.C.; Lettenmaier, D.P. Adjustment of global gridded precipitation for systematic bias. *J. Geophysi. Res. Atmos.* **2003**, *108*. [CrossRef]
51. Morbidelli, R.; Saltalippi, C.; Flammini, A.; Corradini, C.; Wilkinson, S.M.; Fowler, H.J. Influence of temporal data aggregation on trend estimation for intense rainfall. *Adv. Water. Res.* **2018**, *122*, 304–316. [CrossRef]
52. Yang, D.; Ohata, T. A bias-corrected Siberian regional precipitation climatology. *J. Hydrometeorol.* **2001**, *2*, 122–139. [CrossRef]

MDPI
St. Alban-Anlage 66
4052 Basel
Switzerland
Tel. +41 61 683 77 34
Fax +41 61 302 89 18
www.mdpi.com

Remote Sensing Editorial Office
E-mail: remotesensing@mdpi.com
www.mdpi.com/journal/remotesensing

www.ingramcontent.com/pod-product-compliance
Lightning Source LLC
LaVergne TN
LVHW070616100526
838202LV00012B/657